THE EUDAEMONIC PIE

Thomas A. Bass, who holds advanced degrees from the University of Chicago and the University of California at Santa Cruz, was a member of the group whose adventures are chronicled in this book. The author of *Camping with the Prince and Other Tales of Science in Africa* (available in Penguin), he lives in Paris with his wife and daughter.

THE EUDAEMONIC PIE

THOMAS A. BASS

PENGUIN BOOKS

PENGUIN BOOKS
Published by the Penguin Group
Viking Penguin, a division of Penguin Books USA Inc.,
375 Hudson Street, New York, New York 10014, U.S.A.
Penguin Books Ltd, 27 Wrights Lane,
London W8 5TZ, England
Penguin Books Australia Ltd, Ringwood,
Victoria, Australia
Penguin Books Canada Ltd, 10 Alcorn Avenue, Suite 300,
Toronto, Ontario, Canada M4V 3B2
Penguin Books (N.Z.) Ltd, 182–190 Wairau Road,
Auckland 10, New Zealand

Penguin Books Ltd, Registered Offices:
Harmondsworth, Middlesex, England

First published in the United States of America by
Houghton Mifflin Company 1985
Reprinted by arrangement with Houghton Mifflin Company
Published in Penguin Books 1992

1   3   5   7   9   10   8   6   4   2

Copyright © Thomas A. Bass, 1985
All rights reserved

ISBN 0 14 01.6712 9

(CIP data available)

Printed in the United States of America

To the Eudaemons

Acknowledgments

This story belongs to its heroes and heroines. Its strengths come from the patience with which they instructed me in computers, gambling, and the eudaemonic connection; its weaknesses are my own. The manuscript was read entirely or in part by Doyne Farmer, Norman Packard, Letty Belin, Lorna Lyons, Edward Thorp, Tom Ingerson, Ralph Abraham, Ingrid Hoermann, Marianne Walpert, Len Zane, and Jim Crutchfield. I truly appreciate the care they took in getting the facts, and everything else, right.

My thanks go to the following friends who aided this project during the four years required to complete it: Bill Pietz, for supporting the book at a time when it might not have existed without him; Dana Brand, for lending his critical eye to an early version of the tale; and Wendy and Jeremy Strick, for their good company in Paris during the writing of a later version.

Among past and present intercessors at Houghton Mifflin, I would like to acknowledge the aid of Jeffrey Seroy, who knew what this book was about, right from the start; Gerard Van der Leun, who offered wise counsel throughout; Sarah Flynn, who saw it through to the end; and Nan Talese, whose advice and encouragement proved invaluable. For the British edition, my thanks go to Richard Dawkins, friend to eudaemonia, and Michael Rodgers, our sympathetic publisher. Finally, I want to thank Nat Sobel, my agent, and Bonnie Krueger, my wife and best reader.

Contents

Glitter Gulch

As I walk along the Bois de Boulogne
With an independent air
You can hear the girls declare:
He must be a millionaire!
You can hear them sigh and wish to die,
You can see them wink the hopeful eye
At the man who broke the bank at Monte Carlo!

> "The Man Who Broke the
> Bank at Monte Carlo"

We drive into the parking garage behind Benny Binion's Horseshoe Club and circle up the ramp to the third floor.

"We shouldn't be seen talking to each other," Doyne says. "Not even in the street. In case there are any slip-ups, we'll meet later in the Golden Nugget. Why don't you run through the signals again?"

"A bet on red means I take a five-minute walk. Even means sit down and play. A chip on the first twelve numbers and I raise stakes."

This is one of the ways we'll communicate without talking for the next two hours. The other is by computer.

We park the car and lift two pairs of shoes off the rear seat. These are good leather Oxfords with crepe soles. Only on peering inside does one notice that the bottoms are hollowed out. A channel three inches wide and a half inch deep runs from toe to instep. A second cavity is cut into the heel. This is professional work. Uppers and soles have been separated and restitched without a trace.

We reach back for two more shoe boxes. One of them holds our power supplies, known to us as "battery boats" because they look like miniature dories with screw-on lids. The second box holds our computers, which resemble orthopedic insoles with toe clickers built onto the front end. The missing pieces in a jigsaw puzzle, computers and boats fit exactly into the cavities cut out of the shoes. The boats slip prow backward into the heel. The computers snuggle up front under the balls of our feet.

Out of their shoes the components might be mistaken for foot warmers or extraterrestrial tape cassettes. But their beauty lies in what they do: their function is the amazing part.

Molded out of clear casting resin, the battery boats hold eighty turns of hair-thin antenna wire embedded along their outer edge. Built into a circuit inside are a 15-volt battery and four 1.5-volt AAA batteries. From the rear of each boat trails a ribbon cable attached to a model airplane connector. This is a miniature plug with eight pins, each of which corresponds to a different function in the computer — for which the boats act simultaneously as radio receivers, power supplies, and message centers.

Covered with screw-on lids made of polycarbonate "jail glass," the boats have two metal solenoids the size of pencil erasers sticking out of holes cut into the plastic. Activated by a small current, these mechanical thumpers are positioned to vibrate against the heel and arch of the foot. By varying the location and frequency of these buzzes, a computer driving the solenoids can generate dozens of discrete signals.

Doyne and I unscrew the jail glass and load fresh batteries into the boats. "We'll use the carbon batteries," he says. "Our range may be shorter, but they give out less noise."

Packed with batteries, antenna wire, a capacitor, a resistor, two solenoids, and three diodes, the boats are stuffed to the last millimeter.

"Let's power up. Then we'll do a range test and head for the street."

We insert the model airplane connectors into the rear of the computers. Semitranslucent rectangles wrapped in tape — for comfort in walking on top of them — the computers are the brains of the operation. Under the tape they display top and bottom the silver tracings of printed circuits. For the elect who can read these manuscripts illuminated in copper and solder, they represent glistening avenues and piazzas in the great City of Computation. Lying barely revealed beneath the circuits are a host of capacitors, resistors, and diodes, a crystal clock pointing the arrow of time, and dark fortresses of silicon in which reside the powers of language and logic under the control of one pre-eminent chip endowed with memory.

An experienced eye would be surprised by the arrangement of these silicon boxes. The chips governing the computer's two basic functions — logic and memory, volition and destiny — have been loaded separately onto circuit boards, which, in turn, have been folded over on top of each other. Imagine upending Tokyo and fitting its skyscrapers, upside down, into the avenues of New York. You get an elegant solution to a topological problem — and a tight fit. Then imagine running a plastic spacer around the waterfront of Manhattan and filling the island with microcrystalline wax — a petroleum derivative as hard as plastic, except at 300° Fahrenheit, when it flows with the viscosity of molasses. Cool the ingredients back to room temperature and you have a Tokyo–New York computer sandwich hard enough to take a blow from a hammer.

In technical terms, we are slipping into our soles a CMOS 6502 microprocessor with five kilobytes of random-access memory. Apple computers are made with the same chip. We carry another 4000 bytes of memory crafted into a program smart enough to beat roulette at a 44 percent advantage. The program — a set of mathematical equations similar to those used by NASA for landing spaceships on the moon — tracks a ball in orbit around a spinning disk of numbers. During the ten to twenty seconds in which the game is played from beginning to end, the computer calculates coefficients of friction and drag, adjusts for changes in velocity, plots relative positions and trajectories, and then announces where in this heavenly cosmos a roulette ball will likely come to rest on a still-spinning rotor. Its predictive power lies in the fact that the computer in our shoes can play out in microseconds a game that in real life takes a million times longer.

A 44 percent advantage is significantly larger than any other gambling system extant. The payout in roulette is thirty-five to one. For every hundred dollars invested — compounded fifty times an hour — one can expect a tidy hourly return of $2200. The money is sweet, but so too is the glory in beating roulette.

After loading boats and computer sandwiches into our shoes, we cover the equipment with leather insoles into which holes have been punched for the solenoids. There are three buzzers altogether: two on the boat and one forward on the sandwich. Programmed to tickle our feet in three different places at three different frequencies, the solenoids produce a total of nine discrete signals. Our socks, too, have neat holes cut into them.

Inside his left shoe Doyne fits a second battery boat and piece of hardware the same shape but slightly smaller than a computer sandwich. A polycarbonate case filled with inverters, transistors, and a radio transmitter, this is the mode switch. Tapping the clicker that hangs off the front end of the switch drives the computer — via a radio link from shoe to shoe — among various modes, or domains, in its program.

Doyne steps out of the car and stands with his big toes positioned over the microswitches in his left and right shoes. His left toe is expert at motoring the computer among subroutines in its program. His right toe is trained for tapping in data. With Doyne's computer on line and making predictions, another radio link connects it to the computer and solenoids in *my* right shoe. This gives us a three-footed system, with functions divided between data taker and bettor. Since I have no microswitches under my toes, my role is limited to fielding signals radioed from Doyne's computer to mine, and placing bets on the layout. I am the front man of the operation, a foil, a mere interpreter of signs tattooed onto the soles of my feet.

I lace up my shoes and step out of the car. I am walking on five years of labor and several thousand dollars worth of soft- and hardware: a state-of-the-art computer. For all that, the shoes don't do much for my posture. Because of their rigidity, I have to walk in them at a stiff-legged lope. Copper screws, filed to a point, have been mounted on top of the solenoid plungers. The screws allow for a custom fit under the heel and arch of my foot that feels, I imagine, like an application of Vietnamese punji sticks. This latest model in the "Cadillac of roulette systems," as Doyne calls it, is about to get its first road test.

"Let's try for range," he says, walking to the front of the car. "Call off the signals as you get them."

The desert air in November, even at night, is warm enough for us to stand in our shirt-sleeves. Doyne's long face is pinched white under its tan. The skin over his cheekbones is nearly translucent. Thin-lipped, his mouth puckers with concentration. His blue eyes, sunk deep, give him the appearance of looking inward, as if over a landscape or a wiring diagram stretched back from his forehead.

With a mop of blond hair curling over his ears, Doyne looks like a west Texas farmboy about to step out for a night on the town. Dressed in chinos and a long-sleeved cotton shirt loud with too many colors, he leans his six-foot frame against the car. Only on looking closely do I see a hint of shoe leather rippling over his toes. "Did you get that?" he asks.

Headlights swing past us on the ramp. We avert our faces as a car circles up to the next level.

"Three," I say, getting the signal, a high-level buzz on the front solenoid.

"Right. What was that?"

"Nine."

"And this one?"

"Five. Maybe a six."

"Let's hit the street. We haven't driven five hundred miles from the coast to stand around playing electronic footsie."

He reaches into his pocket and pulls out a roll of hundred-dollar bills. "Cash in three or four of these. When I give you the sign to raise stakes, cash in a few more.

"How do you feel?" he asks. "Is it your lucky night for gambling?" His face relaxes into a cockeyed grin. "Give me half an hour to get the parameters set. Stop by the table, and I'll flip you a signal on one of the side bets. If you don't find me in the Sundance, I've moved down to the Golden Gate. And if for some reason I'm not there, look for me in the coffee shop at the Nugget. But watch out. There are two of them. We want the one behind the bar. See you later," he says, loping toward the elevator with the peculiar gait of someone wearing a computer in his shoe.

I walk downstairs and turn right onto Fremont Street, or Glitter Gulch, as it's known, the three blocks of casinos that constitute downtown Las Vegas. Unlike the Strip, whose pleasure palaces are surrounded by greasewood wastes that have to be navigated by car from oasis to oasis, Fremont Street can be managed on foot.

It is lined with the town's older gambling establishments, opulent and faded like the Golden Nugget and the Mint, or just faded, like the Golden Gate and the Horseshoe Club. The street at night is a river of neon, swift and beautiful as it flows down the avenue. Current whines overhead, along with the *pop, pop, pop* of circuits switching on and off. Faces in the crowd turn red, white, and blue. Boys stop and stare open-mouthed. Girls titter. The air is charged. People are juiced on the sheer consumption of it, as if the turbines out at Hoover Dam can be heard throbbing thirty-five miles across the desert.

Examining casinos from Binion's to the Union Plaza, I find them uniformly designed in concentric rings not unlike those Dante passed through at the hem of Virgil. One penetrates first into a dark forest of one-armed bandits patrolled by women wearing aprons full of money. Other women throw themselves into the metal arms of these machines, whose embrace is made friendly to humans by means of pictures — oranges, lemons, and church bells — that spin through windows on their faces. Blue-haired grandmothers dip into Dixie cups of change to feed one, two, three machines at once in a frenzied parody of motherhood. Amid a great din of sirens and gongs and silver trickling down into metal bowls, one hears them coaxing, *Thatta boy. You can do it. Let's have another big one. Wheeeooo! Keep it coming.*

The next smoky circle is reserved for the keno display, the Wheel of Fortune, electronic bridge games, the cashier's cage, and parlors devoted to wagering on sports events. A scoreboard on the wall like those in airline terminals flashes track conditions at Bayview and news of a new filly running at Aqueduct.

The din lessens as the forest opens onto the main floor, which is chandeliered and plush, primary in color and instinct. At this point there is often a threshold to cross, a few steps marking the final descent. Arranged below in circles or squares or one great circle are the kidney-shaped tables reserved for twenty-one. Leaning over what look like giant caskets, men on another part of the floor shake dice and roar for their lucky numbers in the game of craps. A quieter group faces the roulette tables. They shuffle chips onto the layout, stare at the wheel, and sip long at their drinks as the ball orbits around its yet-to-be-chosen number. Farther to the rear in the posher casinos ivory dominoes click in the game of Pai Gow, and bankers and punters in evening dress take turns dealing baccarat cards to each other.

A crowd of spectators watches the big players finger their chips, while other onlookers — dealers, wheel spinners, and croupiers dressed in black bow ties and ruffled shirts — stare with corporate severity from the far side of the green felt barrier. The men in dark suits with thick faces and eye muscles turned to gristle are the pit bosses. Dead center among them, elevated at a little podium, stands the shift boss.

The Eye in the Sky, hidden behind one-way mirrors in the ceiling, constitutes another supervisorial level made up of video cameras and tape loops monitored at a central console. No gesture on the floor escapes the scrutiny of the Eye. The employees below play to it like marionettes. No dealer, croupier, or boss touches chips or money without then clapping his hands and turning them palm upward. No shuffle, cut, deal, roll, or spin is made without the Eye recording it. No player walks into a casino without the Eye remembering where and under what circumstances it last saw that face.

There is a gut rush, an unavoidable jag felt on descending to the main floor of a casino. Cocktail waitresses dressed as bunnies and harem girls teeter through the crowd. The air is charged with sexual cues and cultivated looks of availability. But the color, the spectacle, the precision and formality of it are directed elsewhere, toward pieces of silver, gold, paper, plastic, or whatever else is being used to represent money. A lot of money. Piles of bills with portraits of Madison and Grant on them. Mounds of chips numbered in $25, $100, $500 denominations. Money made liquid as it streams and eddies over the tables like the river of neon shooting down Fremont Street.

With half an hour to burn, I stroll through the Golden Gate, the Nugget, the Mint. I pass the flacks handing out coupons and cocktail waitresses hustling drinks; I stop in the crowd around the roulette wheels. Time and again I watch the ball drop and arc toward its rendezvous with one fortunate number.

Walking to the head of the street, I turn into the Sundance, a second-rate casino, a sawdust joint with the usual mix of slots and craps but less "action." To find the really high rollers you have to look in the carpet joints out on the Strip. But the Sundance tonight has a cherry roulette wheel ripe for picking. The croupier keeps the rotor steady and spins a fast ball up on the track. The wheel should prove no match for computer sandwiches built into magic shoes.

I skirt the floor and walk to the back of the casino. From there I watch Doyne standing at one of the two wheels in play. Positioned at the head of the layout, he doodles in a notebook, looks now and again at the wheel, and then seems to screw up all his courage for the occasional bet on red or black. For a Ph.D. from the University of California, he looks distinctly goofy.

Doyne is passing tonight under the pseudonymous name of Clem from New Mexico. This is a role he first learned as a poker sharp touring the card rooms of Montana, and he plays it to perfection. He appears a half-wit, a mumbler, an innocent soul displaced from the prairie. No one around the table gives him the slightest thought. Roulette players like Clem divide into two general types: those of subnormal intelligence, and system players. Doyne could be either, or both. Las Vegas is crawling with gamblers trying to calculate a mathematical edge over one of its games. The casinos help them by providing pencils, scratchpads, runs of numbers, and diagrams of betting layouts and odds. Doyne standing next to the wheel doodling in his notebook looks like any other do-it-yourself mathematician trying to augur a pattern of numbers where none exists.

In spite of all the books sold on the subject, there is no mathematical system — be it a progression, a betting pattern, martingale, d'Alembert, doubling up, or doubling down — capable of predicting the outcome in roulette or improving the bettor's odds. It is also true that system players, especially those who think they have cracked the missing code, tend to lose money faster than the ordinary stiff who relies on luck. This explains why the casinos are so generous with their free pencils and scratchpads.

It is not by mathematical but by *physical* prediction that one beats the game of roulette. You need to know the exact forces acting on ball and rotor *at each play of the game.* This requires a computer programmed with an algorithm — a general equation describing the physics of roulette — into which you can plug the variables governing the wheel. If the wheel is tilted, you locate the high side and shadow on the track. You calculate the average velocity at which the ball tends to fall off. You compute the rate at which the central rotor decelerates. Given these general parameters — which differ significantly from wheel to wheel — the computer and its algorithm become predictive.

But for this they need more information gathered while the game is in play. This is supplied by a data taker clicking two passes of the rotor in front of a fixed reference point on the frame of the wheel, and two or more passes of the ball in front of the same point. It is now an easy matter for a computer to calculate relative velocities and position, the projected time of fall for the ball, its trajectory over the sloping sides of the wheel, and its final collapse onto the spinning disk of numbers.

As I walk into the Sundance, Clem from New Mexico is engaged in the process of setting parameters. To fit the computer's program to a particular wheel, Doyne carries on a kind of dialogue between his big toes. The microswitch in his left shoe steers the computer into subroutines in its program, while the microswitch in the right shoe clocks the ball and rotor data. A tap routine combining left toe and right toe alters the parameters themselves. To get the algorithm tweaked around to the conditions at hand requires a good eye and split-second reflexes. The process takes anywhere from ten minutes to half an hour.

With five years' practice, Doyne is an ace at driving the computer around its program. He adjusts variables by sight, or from a sixth sense developed by now in his big toes. The remaining variables are fine-tuned by trial and error. Does the ball travel farther than or not as far as predicted? Are there unusual circumstances, such as atmospheric pressure, affecting its behavior? From one play of the game to the next, Doyne notes what the computer predicts against what the ball actually does, until, ideally, the two sets of data could be plotted on top of each other in a bell curve neatly symmetrical about the mean. This laminar hump of data points soaring clear over the x axis translates into our 44 percent advantage, and lots of money. A hundred thousand dollars a month, at our latest estimate.

Once the parameters are adjusted and the computer is clicked into its playing mode, Doyne's left toe takes a break. The right foot can handle the rest, which involves the simple clocking of ball and rotor past a reference point. Doyne at this stage can play the game out of the corner of his eye. With his right toe become an autonomous unit, bouncing over its microswitch like a frog's leg pithed for a demonstration of galvanic electricity, Clem from New Mexico brightens up to chat about the weather and flirt with the hostesses.

I walk to the roulette table and stand behind the players. I see from the croupier's marker that Doyne's green chips are valued at twenty-five cents, the house minimum. Seated on stools along the layout are three other players. In the middle is a large blond woman with a gull-wing hairdo. Her red chips are pegged at fifty cents. Next to her, wearing a Stetson and a string tie, is a gentleman playing with black, one-dollar chips. At the far end of the table is a Filipino in a sharkskin suit. His face obscured in a cloud of cigar smoke, he stands behind a pile of blue chips valued at five dollars apiece.

The croupier gives the rotor a nudge and sets the numbered pockets spinning counterclockwise. He launches the ball in the opposite direction and announces in a flat voice, "Place your bets, ladies and gentlemen." Like a Ouija board player hoping for spiritual intercession, the blonde slides her chips over the baize with her fingertips. She leans back and titters to no one in particular, "Oh, my. I sure could use some luck."

Mr. String Tie hedges his bets on corners and columns and then flips a couple of extra chips onto his lucky number 9. The Filipino, betting late and fast, scatters dozens of chips over the layout in stacks three and four deep. He finishes by mounding a pile of chips over the five-corner bet between 00, 0, 1, 2, and 3 that pays the worst odds on the table.

Doyne places a chip on red. The ball spins with the sound of a marble rolling over a hardwood floor. It drops from the track, arcs neatly between two diamonds, gives a little bounce on hitting the rotor, and then falls into the green cup numbered 00.

"Double zero," calls the croupier, as he covers the winning number on the layout with a glass pyramid.

"Oh, my," says the blonde. "Some people have all the luck."

"Nice play," says Mr. String Tie to the Filipino, who is sucking hard on his cigar.

Using a wooden rake, the croupier clears a pile of losing bets. He sorts and restacks them into a bank of chips stored on the apron behind the wheel. He claps his hands and the pit boss comes over to watch the payout. A square man wearing a crew cut and brown suit, he looks unhappy about the crystal marker and blue chips remaining on the table. The croupier pays out thirty-five to one for bets straight up on 00, and six to one for the corner bets. Using his rake, he slides a pile of chips down the baize to the Filipino.

"Is it my turn next?" asks the blonde, again of no one in particular.

Onlookers pile up behind the winner. They stand as voyeurs witnessing a visitation from Lady Luck.

Doyne places an early bet on red: my signal to take a five-minute walk. I stroll the floor, studying the action. Three craps players dressed in seersucker suits and button-down collars must be in town for a convention. A dealer out of play at a twenty-one table catches my eye and fans her deck face up on the baize.

I walk to the back of the casino and sit in front of the keno board. The machine that blows around the Ping-Pong balls starts up. A pneumatic tube sucks them one at a time out of a glass jar. A houseman reads the winning numbers into a microphone, while another houseman lights up the keno board. A woman chain-smoking Kools, her daughter, and I are the only ones watching.

I flex my toes and take a deep breath before walking back to the roulette wheel. I find the blonde cleaned out. She snaps her purse shut and heaves off her stool. Mr. String Tie, down to his last dozen chips, will soon follow her to the bar. It is a terrible thing to be abandoned by Lady Luck. You go listless. You start apologizing for yourself. You finger your chips without love until, in disgust and resignation, you toss out the last of them without even looking. The Filipino, lighting his second cigar, is holding his own. But the spectators have wandered elsewhere.

Doyne places a bet on even: my signal to play. I sit in the chair vacated by the blonde and hand the croupier three hundred dollars. He claps his hands and the pit boss watches as my bills get stuffed into the cash box with what looks like a wooden meat cleaver. The croupier again claps his hands and shoves across the felt three stacks of red chips valued now, according to the copper disk in front of the bank, at five dollars apiece. The pit boss gives me a good stare.

This is it. The knockover. My debut into the big time. I have the layout in front of me memorized backwards and forwards. I know the arrangement of all the corresponding numbers on the wheel. I have them divided around the circle into octants, eight groups of four or five numbers apiece, that correspond in turn to one of eight different buzzes tattooed by computer onto the bottoms of my feet. I've spent days fielding buzzes and throwing bets onto the layout. I've trained for hours to get my fingers supple around the chips. I've mastered the art of stacking them in my palm and dropping them face up on the baize with no movement in the wrist. I can play by reflex, thoughtlessly, without even glancing at the wheel, cool and fast, while otherwise looking like your everyday mark about to burn up a little discretionary income.

My hair is cut short and styled. I'm dressed in twill pants, nicely tailored, with a sports coat, cravat, and shirt opened two buttons down the chest. For small talk, in case anyone asks, I own a restaurant in Capitola, California. Part owner, actually. French. Entrées around fifteen dollars. Specialties of the house ranging from *moules marinières* to *boeuf bourguignon*. I'm in town for a toot. A couple days off before the holiday season picks up.

The cocktail waitress taps my shoulder. "A drink on the house?" she asks.

The croupier flips the ball up on the track. "Place your bets, ladies and gentlemen," he says, without any ladies present.

Mr. String Tie pushes his last chip onto number 9 and stands up for a stretch. The Filipino extends his pinky ring over the layout and covers the green felt with a rash of chips scattered on corner bets, column bets, and numbers split twelve, six, and three ways. It looks as if he has more money on the table than any win could recoup. Doyne places a twenty-five-cent bet on black.

The ball whirls smoothly around the track and slows for its final revolutions. The cups below spin successively red, black, and green. I wait for Doyne to enter data and transmit a prediction from his computer to mine. Like time machines speeding up the present, our computers are going to peer into the future and chart the trajectory of the game a crucial few seconds in advance of its being played. I get a high-frequency buzz on the front solenoid. A three. The third octant. Including numbers 1, 13, 24 and 36. I stretch over the baize and cover the first three numbers with chips. I skip the 36 at the bottom of the layout and substitute instead the 00, which lies near it on the wheel and closer to my seat.

Like a basketball player watching a free throw sail up and into the basket, I lean back on my heels and wait. I turn to the cocktail waitress and order a Tequila Sunrise. I watch the Filipino puff his cigar. I smile at the pit boss. I'm not even looking as the croupier calls out the number 13 and places his pyramid on top of my bet. *Why would anyone play roulette*, I think to myself, *without wearing a computer in his shoe?*

I

Silver City

Prediction is very difficult, especially of the future.

Niels Bohr

Down in the red desert country of southwestern New Mexico, Silver City is famous, or at least notorious, on several counts. Geronimo lay low in the nearby mountains while Billy the Kid shot his first of many men. Herbert Hoover, fresh out of Stanford, got his start in Silver City as a mining engineer. Fifty years later, the struggle of workers and their families at Empire Zinc was featured in Herbert Biberman's classic film *Salt of the Earth*. Violence of a different sort, abstracted from cowboys and Indians, confounded local residents on the morning of July 16, 1945. They woke to the blast and peered through rattling windows to see the glow of the world's first atomic bomb, exploded two hundred miles to the north on the lava beds of the Jornada del Muerto.

On the southern edge of the Gila Wilderness, Silver City straddles the threshold, at six thousand feet, between forest and desert. The Continental Divide, after wandering through the Black Range of the Mogollon Mountains (pronounced *muggy*-OWN), zips through town on its way headed due south into the Sonoran Desert. A city of twelve thousand souls, the seat of Grant County and biggest way station for a hundred miles in any direction, Silver — as the residents call it — is pretty much in the middle of nowhere.

A thousand years ago the Mimbreño Indians wandered into this pleasant stretch of upland desert and called it home. The land had much to recommend it. Forested with Ponderosa pine and Douglas fir, the ten-thousand-foot peaks of the Mogollons gave rise to the three forks of the Gila River, along which the Mimbreño built cave-side dwellings and painted frescoes of a landscape rich in arable fields and wild game. The Indians farmed the valleys and chased the game south into countryside that tipped from xeric woodland — filled with scrub oak, mountain mahogany, juniper, and piñon — into bushier terrain of creosote, jumping cholla, and yucca, before dropping finally into a howling desert of dry playas and alkali flats. At the base of the mountains lay a favorite camp for these hunting parties, a stretch of springs and native prairie that the Spanish, on their belated arrival, named La Cienaga de San Vicente, the Marsh of St. Vincent.

The affixing of saintly place names accompanied more serious incursions by the Spanish into the Gila Wilderness. They sold the Mimbreño into slavery and ordered all resisters killed in a war of extermination. A Spanish officer engaged in this civilizing mission discovered what the Indians had long known about the area's mineral wealth — that copper could be harvested here from underground veins as thick as ferns. He returned in 1804 to open the Santa Rita copper mine, an original act imitated by successive waves of hard-rock miners up from Mexico, forty-niners out from Boston, gold rushers down from Leadville, and a later band of enthusiasts whose metallic fever ran so high that in 1870 they rechristened St. Vincent's marsh with the more promising name of Silver City.

The town has faded and got a bit ragged around the edges with shopping plazas and subdivisions, but much of it still appears as it did in its heyday a hundred years ago. Built higgledy-piggledy on hilltops and along elm-lined streets are the brick homes of old miners who hit a vein and subsequently called themselves bankers. Imposing structures, with Victorian porches, mansard roofs, Gothic turrets, and widow's walks, these houses command vistas across the desert scrub to the flanks of Geronimo Mountain.

Looking from the second-story windows of these houses, one spies on every horizon the presence of minerals. Due east, under a monolith known as the Kneeling Nun, stretches the mile-long pit of the Santa Rita copper mine. Its ore — exploited multinationally right from the start — was originally transported four hundred miles by mule train into Chihuahua. Now, mined by Kennecott and Mitsubishi, the shipment goes to Japan. The twin stacks of the mine's smelter rise over the nearby town of Hurley, and to the north squat the equally dusty houses of Hanover. In Silver City itself the hills are littered with mine shafts and tailings from abandoned veins, while directly behind the County Courthouse a still-active manganese mine chews away at the face of Bear Mountain.

Out beyond the scraping and digging on which its fortunes were founded, the view from the high ground in Silver City gives way to unpeopled prairie. This is welcome ground to big dreams. Out of it spring self-reliant souls. An old part of the country, discovered and settled long before the Pilgrims disembarked from the *Mayflower*, it is also among the newest, with the feeling of being a frontier, a border territory, unsettled and wide open to chance. Here one finds the last avatars of the American mythos — cowboys and Indians, hard-rock miners and barroom confidantes — with room enough still for them to stretch out and dream the old dreams of freedom and independence.

Born in Houston, Texas, in 1952, James Doyne Farmer, who goes by the second of his two given names, which is pronounced like a variant tone row, *Dō-an*, was six years old when he and his family moved to Silver City. They settled into a mild neighborhood spiced with Mexicans and college students while James Doyne Farmer, Sr., who goes by the *first* of his two given names, reported to work as an engineer at the Santa Rita mine. Doyne by then had already read his first novel, *Robinson Crusoe*, and over the next few years Mrs. Lynch, the town librarian, would lead him in an attack on the rest of world literature. During her summer reading contests, the number of stars he acquired for the consumption of texts — from Dostoyevsky through Hemingway and Huxley to Tolstoy — was second only to Kitty Kelley's.

"Then in sixth grade," Doyne told me one day as we were driving down Las Vegas Boulevard on our way to play roulette, "I pooped out. As a kid I was plump, had a fierce temper, and didn't get along well with other people. I decided I had to work on my personality."

The turning point had arrived earlier that summer. Suffering as much from the cure as the disease, Doyne was confined to bed with rheumatic fever for months of total inactivity.

"I became very depressed. When no one came to visit, I decided there must be something wrong with me. I went through a complete self-reappraisal and made certain vows. When the teacher asked a question, I would no longer raise my hand. I would do whatever I needed to become popular. From being loud and boisterous I became quiet and shy. I spent sixth grade learning the technique, and seventh and eighth grades perfecting it. By ninth grade I was back to some kind of balance, but in the meantime I had made a lot of friends."

In sixth grade Doyne also committed himself to pacifism, although this project fared less well. It met insurmountable obstacles down at the paper shack, where he reported every morning at five o'clock to fold the day's delivery of *El Paso Times*es. At that hour the only people awake in Silver City are the paper boys, Mr. Shadel the baker, the police (who are actually fast asleep in their patrol car), and the female boarders over at Millie's, a Victorian structure on Hudson Street that was reputed, until its recent closing, to be the best whorehouse in New Mexico.

Since Route 180, the main road through Silver City, runs from nowhere in the north to pretty much nowhere in the south, anyone stopping at 5:00 A.M. to ask directions from a paper boy was most likely looking for Millie's. It was only two blocks from the paper shack, but a quirk of local geography made the town confusing to visitors at that hour of the morning. The marsh on which Silver City had been built provided an oasis on the edge of the desert. But every summer when it rained, flash floods would rush out of the Mogollon Mountains and rip down the middle of Main Street.

Forced to abandon this misplaced thoroughfare, the town watched it metamorphose into something called the Big Ditch. The Ditch today is a canyon scraped down to bedrock sixty feet below those buildings that have yet to collapse into it. "For many years," reported the local newspaper, "the Ditch has been the literal 'jumping off place' for petty thieves and miscreants who were running from the law, and many a wretched 'wino' has sought the privacy of its dark green shadows." Millie presided over the eastern, less savory side of the chasm. In that direction lay Madame Brewer the witch's house, the barrio, the dump, and the desert. Not that it was much better on the other side of the Ditch. Here too were Mexican adobes and desert scrub, although perched above them on a hill was the campus of Western New Mexico University.

The only "good" part of town lay to the north in Silver Heights. But none of the characters in this story comes from that part of town.

At 5:00 A.M., when not directing traffic around the Big Ditch, Doyne reported to the paper shack for his daily lesson in Satyagraha. The other paper boys fancied themselves tough customers, the two orneriest being James Wetsel, who tooled around town in a car he called his "pussy wagon," and Herbie Watkins, who was Wetsel's goon.

"Their big thing was to beat the daylights out of me every morning," Doyne said. "I'd walk in and one of them would tackle me. They'd take off all my clothes and throw them outside in the snow. Being a pacifist at the time, whenever they did anything I'd go limp. I had read something in the sixth grade that convinced me that was the best response to violence. But it didn't work very well."

It was also in that busy year that Doyne encountered the most significant influence in his life. He was attending the weekly Boy Scout meeting when someone stood up and introduced himself as Tom Ingerson, a physics teacher at the university who wanted to help out with the troop.

"I homed in on him immediately," said Doyne. "I didn't know at the time what a physicist was, but I knew it was some kind of scientist, and that that's what I was going to be."

After the Scout meeting, walking home with Ingerson, Doyne told him how he wanted to be a physicist. The two of them — a twelve-year-old kid and a twenty-five-year-old college teacher fresh out of graduate school — recognized an immediate attraction for each other: the gravitational pull of minds equally restive. They began what Ingerson described as "hundreds and hundreds of discussions about everything under the sun. In a town where the average random kid thought only about going to the mines or becoming a shopkeeper, Doyne and I were bored for much the same reasons."

Something about this man intrigued Doyne, and on later visiting Tom's house, he got an intimation of what it was. "Filled with a mess of books and tools and Salvation Army furniture, the house was completely disordered inside and out, and somehow this really impressed me."

Tom Ingerson's anomalous presence in Silver City was due to a mixture of romance and political miscalculation. On finishing high school in El Paso, Texas, Ingerson, another son of an engineer, spent an alienated four years at Berkeley before going on to do graduate work in physics at the University of Colorado. While nearing the end of a dissertation on Einstein's cosmology and the theory of general relativity, he went looking for a job. This was 1964. Sputnik and the cold war had the country crazed for scientists. They could snap their fingers and walk into laboratories anywhere from Boeing to Bell. Colorado was a good school, Ingerson a bright fellow. He should have gone places. Instead, he went nowhere.

His letters remained unanswered. No one interviewed him. In his naiveté, he took the matter personally and got depressed. It was only years later that a friend at Motorola showed him the bad news in his file. A strong-willed loner like Ingerson may have been ignorant, or he may even have courted disapprobation by naming Frank Oppenheimer as a reference. But even at that late date, the reputation of Robert Oppenheimer's brother as a fellow traveler made him leprous company. Having him recommend you to a prospective employer was like announcing a contagious disease.

So Ingerson found himself banished to the hinterlands of southwestern New Mexico. His one and only job offer came late in the season from Western New Mexico University, the old Territorial Normal School, which, in spite of its new name, was still primarily a teacher's college. Taking up his duties at WNMU as the sole member of the physics "department," Ingerson consoled himself with a number of thoughts. Roughened into mesas and wilderness tracts, this was country he already knew and loved. Once out in the middle of it, he could keep himself amused. He devised a long list of projects and schemes, the most romantic of which involved a gold mine in the Jefferson Davis Mountains of west Texas. A sixteenth-century Spanish mine, the claim had come into Ingerson's possession via his uncle Jim. This grizzled prospector, with a degree from the yarn-spinning school of life, had told his nephew a story of fabulous riches. No less than nineteen tons of gold had supposedly been saved from ambush and buried somewhere under his mountain.

Ingerson's other uncle, Earl, a geologist at the University of

Texas, called the story hogwash, but the newly hooded Doctor of Physics believed strongly enough in his golden legacy to describe it as "the reason I moved to Silver City in the first place."

Ingerson is the quintessential physicist. Possessed of a resonant, slightly didactic voice, he can discourse at length on any question pertaining to matter and motion. There is not a mechanical, electrical, computational, or cosmological problem toward which he has not directed his powers of ratiocination. The simplest inquiry garners his *total* attention. An offhand remark about differences in color film, for instance, will spark him into delivering a minilecture on spectroscopy and the psychology of color perception. During one such discussion, interrupted for twelve hours by other matters, Ingerson resumed his comments at precisely the point where they had broken off. "That's just the way I am," he said. "I like to complete a line of thought."

Ingerson's blue eyes shine out of a broad face that reddens easily with humor. But from his domed forehead down to his track shoes, his entire demeanor bears the mark of being dictated by rationality. Compact and solid, his body looks as if it might have been designed according to sound energetic principles. Ingerson dresses in layers of long- and short-sleeved cotton shirts that can be seasonally adjusted, and no social occasion merits the slightest change in attire. He travels with backpack and bedroll and recommends, when pressed for space, the omission of all inessentials, such as toothbrush handles and toothpaste.

"I'm a physicist," he said, "because every other way of looking at the world is too difficult for me. In physics we abstract things into simple systems, and if the world doesn't fit, we just lop some of it off and get it simple enough for our models. This is really very easy to do, even though most people think physics is hard."

The night he and Doyne first met and walked home together from the Boy Scout meeting, they talked about Ingerson's gold mine and how, with as little as seventeen million dollars from it, one could build a rocket and travel to Mars. "Tom thought the space program was being utterly mismanaged," said Doyne. "Old farts like Wernher von Braun were making stupid rockets, but he knew the way to do it more cheaply and efficiently." While still in high school, with chemicals ordered from advertisements in the back of *Popular Science*, Ingerson had built and fired dozens of rockets for launches of up to five miles across the desert. Later, in college, he had worked summers at the White Sands Missile Range testing larger rockets. "You have to remember," said Doyne, "that Tom knew what he was talking about."

Doyne soon found himself enrolled as the charter member in what would become Ingerson's major Silver City diversion. One day the young college teacher declared his residence open as Explorer Post 114. A duplex stationed under a clump of Chinese elms lining a tributary to the Big Ditch, Ingerson's house was quickly overrun with boys soldering radios, practicing Morse code, stripping down dirt bikes, and tuning the engine on a Dodge van. Christened the Blue Bus, this vehicle would carry the peripatetic Ingerson and his extended family thousands of miles, from Alaska to the Andes.

Their first trip was to Boulder, Colorado, where the Explorers accompanied Ingerson for the oral defense of his dissertation. Once bitten with wanderlust, they spent every free moment after that touring the Gila Wilderness and Sonoran Desert. At Christmas they traveled to Mexico and during the summer made longer trips to the Yucatán, Panama, and Peru. When not on the road, they spent their time in Silver City raising money at auctions, turkey shoots, car washes, demolition derbies, and a yearly fair known as Gold Rush Days, in which the Explorers re-created a town full of Indians, assayers, sheriffs, prospectors, and claim jumpers searching for hidden caches of golden rocks. "My personality is basically synergistic," said Ingerson. "I don't do much by myself. But together with someone else, things happen. My greatest pleasure in life is seeing other people have a good time."

Besides travel, Explorer Post 114 specialized in electronic tinkering. Ingerson attracted boys of a scientific bent, or perhaps he sparked that interest in the first place. "I've always been perturbed by my influence on the kids," he said. "I never set out to make them physicists. I didn't care what they became. I enjoyed their company and thought I might help make them better adults than they might otherwise have become."

The troop acquired other of its leader's characteristics, including a disdain for social approbation. "As long as the Post lasted," said Doyne, "we prided ourselves that not one person earned a single merit badge or advanced any rank in scouting." When Johnny Reynolds said he was considering getting the one additional badge required to become a Life Scout, he was dissuaded by a combination of physical arguments known as pink bellies, kidney pulses, and bellybutton tests.

After Doyne bought a Honda 50, and Tom a Yamaha 100, motorcycles became another big part of the Post's activities. "My house, dump that it was," recalled Ingerson, "was made even dumpier by having motorcycle engines all over the living room and crankcases in the kitchen sink and pots of grease and oil everywhere. It was terrible." While out in the driveway laboring over inert metal, Ingerson and the boys conversed for hours about motorcycles and life. "We talked," said Doyne, "about physics, sexual mores, history, politics, everything."

"I treat all questions as fair," said Ingerson, "even if I have to say that no one knows the answer. Asking questions about the number of stars in our galaxy or the origins of the universe, these kids had parents and teachers ignorant of whether or not *anyone* knew the answers."

In his first year of high school Doyne took over the spare bedroom in Ingerson's house. His family, which now included a younger brother, was leaving Silver City for Peru, where his father would work as a mining engineer. Moving again from the Peruvian desert to an iron mine in the jungles of Venezuela, the Farmers would spend the next seven years in South America. Doyne visited during vacations, and sometimes worked in the mines as an *obrero*, but he lived mainly with Tom during his first two years in high school.

"I never regarded ours as an adult-child relationship," said Ingerson. "Doyne has always been just about my closest friend. It was like we were roommates. I've always felt more like Doyne's older brother. I never made any rules. There were never any fatherly disciplinary talks. He always seemed to me to be grown up. Doyne was Doyne and he did what he thought best."

He may have been the smartest kid in his class, but Doyne still got in trouble with teachers who resented his obvious boredom with school. "At Silver High you had to be dumb to get a B. The C's were reserved for kids who couldn't speak English. That meant that anyone with half a brain who spoke English got all A's."

Doyne had other interests he cared more about than school. He played viola and guitar and later took up the blues harp. He passed the test for a ham radio license and worked on science projects with Tom, including an unsuccessful effort to build a facsimile machine for sending pictures via radio. He acted in school plays and got a reputation for being a daredevil. At a chicken-eating contest held during the all-you-can-eat night at the Holiday Inn, he won handily by consuming thirty-two pieces of fried chicken. "And I'm pretty sure we went out afterwards for dessert."

A lanky kid shooting up through adolescence, Doyne had a big appetite as well for adventure, books, friends — everything his mind could come in contact with and absorb. He also had a knack for making his enthusiasms the general rage. "We looked up to Doyne and his crowd," said a former schoolmate. "While they went off on adventures to Alaska and South America, we girls in town didn't have anything comparable to do. We took over the school paper or got silly instead. Doyne unconsciously broke a lot of hearts in Silver City."

Explorer Post 114 met weekly on Wednesday nights for a game of capture the flag and lecture on a subject of general interest. The address one week was given by a twelve-year-old named Norman Packard. Discussing crystal radios, their design and construction, he spoke in a barely audible squeak. Possessed of an outsized head perched on a mere spindle of a body, Norman, if you wanted to know about radios and electronics in general, was the person to ask.

Two years shy of the official age for becoming an Explorer, he was adopted into the Post anyway. This was Doyne's first conscious meeting and the start of a lasting friendship with his younger protégé. I say *conscious* because in Silver City everyone knows everyone else all along.

"I have a feeling from looking at pictures of myself growing up," said Norman, "That my head was always too big for my body. It accentuated the cerebral quality that was a social stigma in the first place." On the way to sprouting past six feet two inches in height, Norman remembered "growing a little bit and then having to figure out where my hand was. My father called me Slewfoot, just to remind me that I didn't really have it all together."

The first of six children, Norman Harry Packard was born in 1954 in Billings, Montana, where his father managed the Sears, Roebuck store, "although he didn't fit too well into the corporate

mold," said Norman. With the promise of a new career, Mr. Packard was lured to Silver City, where his wife's parents had built a
chain of five-and-ten-cent stores into more extensive holdings. "My
grandfather at the time owned half of Silver City's lower- and middle-class housing. My father was supposed to break into the landlord business by managing apartments, but it wasn't his cup of
tea."

Norman's father and mother simultaneously launched themselves into yet another career. They went back to school, got their
teaching certificates, and moved the family south to Hachita, a
town near the Mexican border with a population of seventy-two,
including all eight Packards. Spread over a great plateau of desert
wedged between the Hatchet and Cedar mountains, this is ranch-
ing country. "It was pretty isolated," said Norman, "although it
was nice to grow up with so much space. Our front yard consisted
of a range of mountains with more vacant miles than you could
think of walking."

By the time Norman gave his first Explorer Post lecture, the family had moved back to Silver City, where his mother taught second
grade and his father handled junior high school mathematics. The
Packards occupied a brick house on the corner of Bayard and
Broadway, the old commercial street in town. Theirs was the largest private residence in Silver City, although its twenty-three
rooms had suffered considerable depredation, including subdivi-
sion into numerous apartments. The growing Packard family re-
claimed most of these, although a laissez-faire policy left several
boarders in place until they died or moved off. In a house so big
that its rooms were referred to by number, an extra person or two
threading up and down the front stairs was barely noticed. Doyne
himself, during his last year in high school, lived with the
Packards.

Directly across from the Packard house lay the Post Office, until
it was moved a few years ago to a more promising part of town.
While the foot traffic was still good, Norman took advantage of the
location to open two of his three successful businesses. On an enclosed porch facing Broadway, he operated the largest tropical fish
outlet in Grant County. Posted hours for the Silver Aquarium were
"after school," when Norman was also open for business as a TV
repairman. While still in the seventh grade, he had convinced Col-

by's hardware store to hire him as the assistant TV repairman. He was supposed to help the man nominally in charge of the operation, but it was really Norman who fixed the TVs. Quitting after a year and a half, when Mr. Colby refused to raise his salary above $1.50 an hour, Norman turned his bedroom into an electronics shop complete with multiple power outlets and a workbench covered with cannibalized TVs.

Norman by then was also reporting daily to the paper shack, where James Wetsel and the other tough customers had been replaced by a more sober crowd. Norman evaluates the success of his early business careers as follows: "I was a rotten paper boy — that is, rotten at making money. It was my first experience with a monetary scheme that looked good in theory but that actually proved more dismal. I was supposed to make forty dollars a week, but never collected anywhere near that amount. I enjoyed the tropical fish, and probably sold enough gravel and fish food to break even on personal expenses. The TV repair business was more lucrative and pretty simple, except for the occasional glitch that just wouldn't go away."

Norman had introduced himself to Tom Ingerson by asking questions about a radio he was building. But long before he met Tom, Norman knew that physics was his métier. "Since the second grade it's been clear that I was going to be a scientist. That was my burning interest. I remember a student teacher keeping me after class in the fourth grade to ask me what I was going to be when I grew up, and I told her without a moment's hesitation, 'A nuclear physicist.' I was on the track to being a physicist before I met Tom, but it was helpful having him there to cultivate that side of my existence. From the perspective of a teenage kid, Tom knew everything there was to know about physics."

Ingerson lasted four years in Silver City. His primary motive for staying that long, he said, was to take care of his Explorers. He subsequently moved to the University of Idaho, which had fellow members in a real physics department. But he left behind in Silver City the germ of an idea that would soon blossom in unexpected ways.

Ingerson is a complex man. He speaks about himself with such transparency that one forgets he is talking about contradictions.

"As a physicist, I'm yin-yang," he said. "I have the practical knowledge to do things. Knowledge in computers, electronics, telescopy, mechanics. I can build things and get them to do what I want. But on the other hand I'm basically a theoretical physicist who likes playing with ideas. But not to the exclusion of other things in life. Ideas by themselves bore me. If it came to choosing between ideas and people, I could do without theoretical physics."

Ingerson embodies almost to the point of caricature what it is to be a rational human being. He speaks in streams of calculations, variables, and weighted options. His emotions are quantized and assigned state-of-the-world probabilities. In a letter written to Doyne several years after leaving Silver City, Ingerson speculated about his west Texas gold mine: "I don't believe there is 10^8 worth of gold in that hole. I would be willing to believe 5×10^6. With luck and proper organization, we would be able to get $\frac{1}{5}$ of that out of the various governments. Thus, one might expect that there is a chance of making a million dollars out of it." A less rigorous mind might have *guessed* a million dollars. Ingerson *derived* it.

If he led his charges into studying physics, he also instilled in them a healthy distrust for the profession. Ingerson's own inclinations as a rebellious loner, combined with poor treatment at the hands of academic mandarins, gave him a restlessness, both intellectual and physical, that kept him on the move from idea to idea and place to place. He junketed thousands of miles in the Blue Bus and then farther afield to Europe and Russia. He shoved off overland to Chile for a year's sabbatical, and then struck out from there to New Zealand, Thailand, Nepal, and on around the globe.

As Ingerson worked his way up the ranks of the tenured professoriat, newspapers and magazines began to quote him as an expert on planetary syzygies and other astronomical phenomena. He made himself indispensable to his colleagues as the technician who could build anything they needed from rockets to solar collectors. At the brute level of manipulating technical ideas and machines, there are few people more at home in the twentieth century than Ingerson.

While thinking of ways to free himself from teaching and strike out on his own, the problem always resolved itself into one of financial independence. "Money is the key to freedom," he told

Doyne and Norman. "There are two ways to make it, capitalism and theft." The odds on getting caught at theft were too high. That left capitalism. Ingerson imagined founding a company, a kind of capitalist's alembic in which he and his friends would transform ideas into gold. It would exploit inventions, concepts, and new modes of production — the ideas for which Ingerson possessed in abundance.

After researching sources of funds, fiscal control, laws of incorporation, taxes, and patents, he generated a list of ideas to be exploited by his company. First, there was the family gold mine. Then, in no particular order, came rockets to Mars, digital high-fidelity amplifiers, slow-scan TV transmitted via radio waves, computer software for programming Isaac Asimov's "positronic brain," desk-top laser libraries, "system independent" houses built underground, computerized micromaps, dirigibles, and weightless cubes. "I have a notebook full of pretty damn good ideas," he wrote to Doyne from Chile. "Some eminently exploitable."

Ingerson and his young friends talked for hours about turning inventions into money. If they started a company, how would they handle the division of labor? What kind of organization could function both democratically and efficiently? Ingerson imagined this enterprise formed around a community, an anti-entropic center resistant to "the fissioning pressures of society. I feel that man is a tribal unit," he said. "We do better in families than in megalopolises." He wanted to found his company in a place removed and self-sufficient, but still accessible to the flow of technology and ideas that Ingerson, as a "high technocrat," thought to be, at this perilous stage in our existence, necessary to survival.

As he wrote to Doyne in the mid-1970s: "Doom cryers have existed before in all times and places. I shouldn't be too pessimistic, but I certainly am not very optimistic about the future of civilization. I am beginning to suspect that the future may regard the sixties as the point where the world peaked out. We have been running a civilization based on the assumption of infinite supplies of cheap energy, cheap raw materials, and space on the planet on which to put more people. All of these assumptions are clearly straining at the moment. This in itself is not that bad, since I think that technology has the means to fix things. But I don't think that the political process in any country in the world is in a position to take advantage of technology to produce a fix."

He also wrote, "The world is going to collapse around us in a few years, and when it does, I would like to be able to feel in my own mind that I had at least *tried* to stop it, even if I fail, which I surely will."

Ingerson understood that if nothing else, he was offering his friends a challenge. What would *they* do in the face of these contradictions? "I am more and more coming to realize what it is to be a dreamer. We love to think about things, dream, imagine. But when it gets down to the nitty-gritty, a real act of doing something major is too much trouble, because anything worth doing carries with it tons of shit which is no fun to do. A little kid loves to think about neat projects, and it doesn't matter much to him that he doesn't get anything done. An adult eventually loses that, which is really a crime, because of the pure practical problem of earning a living, until he finally loses his dreaming altogether and becomes an old fogey.

"I have been very lucky on that score," Ingerson concluded, "for I have fallen into a job that allows me to behave very much like a little kid, because the university pays me enough so that I can afford to get nowhere. The result is that I attract kids around me who, like me, want to stay a little kid, but so far none of them has figured out how."

When Ingerson moved to Idaho, Explorer Post 114 did less traveling and more racing of motorcycles. But it was time for all of them to think about moving on. Straddling Idaho's western border with Washington, the University of Idaho dominates the town of Moscow, an outpost in the Clearwater Mountains halfway between Spokane and Walla Walla. Driving past the Tetons and up over the Bitterroot Range to the headwaters of the Snake, Ingerson had reached Moscow by steering the Blue Bus twelve hundred miles north along the Continental Divide. Doyne thought it a good trail to follow.

As a high school junior he had been admitted early to the University of Idaho, although his mother, thinking he needed to acquire some "polish," convinced him instead to attend a Catholic boarding school in Tulsa, Oklahoma. Doyne lasted four days in Tulsa before flying to Moscow, where he moved into Tom's attic and started college.

But his mother was right. Doyne had no polish. He farted loudly in public and wolfed down large quantities of food. As a dinner guest at someone's house he once consumed twelve pork chops and a bowl of scraps reserved for the dog. What he had instead of polish was a wide-ranging mind. He got turned on to ideas and pursued them with messianic ardor. The motor behind this restless energy, as a friend described it, was "a fear of boredom, of not reaching out and grabbing everything there is in life, reaching it and touching it and tasting it, a fear of skimming along the surface and waking up years or decades later to regret something not done."

At Idaho Doyne for the first time formally studied physics. The course that year happened to be taught by Ingerson. Simultaneously enrolled for his senior year at the local high school, Doyne found it "a schizophrenic life. I started growing my hair long and smoking marijuana and dating girls from both high school and college. It was a big year for dates. These were your classic affairs with a trip to the movies or a dance, and parking afterwards in the Blue Bus. But I made the mistake of falling in love with a Mormon girl, so that getting even a kiss took hours of maneuvering."

The Bus, a Dodge Sportsman van with a slant-6 engine approaching its second turn around the odometer, had been structurally modified so that it looked, at this stage in its life, like a cross between an artist's atelier and a dump truck. A second story had been added to the roof and a kitchen built onto the back. The inside had been fitted with bunk beds, a carpeted platform, and storage compartments. The Bus slept eight comfortably, which it did once for two months, although on one record-breaking occasion no fewer than thirty-seven Boy Scouts had squeezed inside.

In 1969, with social solidarity afoot in the country, even someplace as removed as Moscow, Idaho, voted no to the Vietnam War by blowing up the ROTC building. "I knew it was lightweight, anyway," said Doyne. "After a year in Idaho I wanted to get out and hit the big time, go where things were *really* moving and shaking. I wanted to meet degenerate drug-crazed hippies and intellectuals. I also realized I had to get away from Tom. Move out on my own and develop my own identity." Ingerson had appeared in Doyne's life as father, brother, teacher, and friend. So one can imagine why the emotional currents ran deep.

That summer Doyne bought a Ducati 250 for seventy-five dollars and rebuilt the engine. Using a pair of hiking boots as "Chinese foot brakes," he biked in ten days from Idaho to Los Angeles, where he caught a plane for Venezuela. He spent the summer working in his father's iron mine and returned that fall to begin his first year as an undergraduate, with sophomore standing, at Stanford University.

Of all the places in the world to look for drug-crazed hippies and intellectuals, the San Francisco Bay Area in 1970 was undoubtedly the best. "I spent my sophomore year smoking dope, meeting lots of people, chasing women around, being existential, frustrated, depressed, going through a big identity crisis, growing my hair." Doyne's other diversion that year was the blues harp, which he played in a local rock band.

"I screwed up sufficiently my first year at Stanford that I was put on academic probation and considered quitting. A friend named Dan Browne and I were thinking of moving to San Francisco and opening a smoothie stand, or going into the house painting business. Then we thought of motorcycle smuggling."

On arriving at Stanford, Doyne had moved into Jordan House, the old Delta Delta Delta sorority, which the university had leased to "a bunch of aspiring hippies. There were thirty-five of us in there in a cooperative living arrangement. You could call it a commune. We did our own cooking and managed our own business affairs." When it closed for the summer, the Jordan House community shifted over to its sister establishment, Ecology House. To save on rent, Doyne and Dan Browne camped out back in the parking lot and set to work on the first phase of motorcycle smuggling: the rebuilding of a BSA 250. Doyne drove the bike later that summer to Guadalajara, where it blew a rod and was traded for a dysfunctional Vincent Black Shadow, which was crated up and shipped to Norman in Juárez. So much for motorcycle smuggling.

Before buzzing off to Mexico, Doyne was rebuilding the gearbox on the BSA when someone came over to introduce herself as a former resident of Jordan House, now returned to Stanford after a year out of school working in New York City. The name Letty Belin was already familiar to him. He had first seen it on cleaning out a closet and discovering a pile of homework for a course on computer programming. The homework was Letty's. It was marked with straight A's. Doyne had heard other stories about this brainy fair-haired girl, and she too had been informed about the wild guy camped in the parking lot. "It was just one of those times," said Letty about their first meeting, "when everything absolutely clicked right from the beginning. I have a pretty misty-eyed vision of the whole summer." An aspiring motorcycle smuggler and a daughter of the Boston aristocracy falling in love with each other in the parking lot of Ecology House . . . Those were unusual times.

It was Letty more than Doyne who looked native to the golden West. Tall, with a boyish figure and straight blond hair framing a classically proportioned face and sensuous mouth, she moved through life with presumptive ease. The A's on her computer homework were merely part of a long string of accomplishments that ranged from calf-roping prizes in summer camp to a Phi Beta Kappa key. Gracious to a fault, if her new friend had no polish, Letty had plenty to spare.

Usually dressed in blue jeans, Oxford cloth shirts, and running shoes, Letty managed to look Cos Cob trim, yet slightly frayed, like a once-tended lawn reverting back to its natural state. Born in 1951, the youngest of five children, a product of the Shady Hill School in Cambridge and St. Timothy's outside Baltimore, coming from a family of lawyers, educators, and servants high in the realms of public policy, Alletta d'Andelot Belin was trying to get away from what she called her "easternness."

She and Doyne made a physically striking couple. But they also cared about community and politics and leading the examined life, for which they were willing to jettison the old values of class and prejudice, as well as the old models for sexual division. Everything was up for grabs, to be thought out item for item in searching for new forms of social organization. Hoping to engage their talents doing something nervy and smart, they were looking for their break.

Instead of dropping out of school, Doyne moved back into Jordan House with Letty and spent his junior and senior years at Stanford "getting serious. I was going to give physics one quarter of trying hard to see how well I did. Either I made A's in all my courses, or I was going to quit." Declaring himself a physics major this late in his career meant that Doyne had to take five required science courses every quarter. This was the toughest major on campus. While grinding through classes in quantum mechanics and statistics, he found them "incredibly competitive. No one would talk to you about the homework assignments, and no one asked any ques-

tions in class. It wasn't the cool thing to do. Once again I felt schizophrenic. I was living in a commune, but taking all my courses
with science jocks. I was hanging out with artists at night and
nerds by day."

In 1973 Doyne graduated from Stanford wearing a gorilla costume. This was done in punning deference to the Viet Cong, but
the *San Francisco Chronicle* missed the point. They pictured him
on the front page as an example of student frivolity. "I'm a Yippie,"
he later told me. "My global strategy is to bring it all down. I think
it's impossible to work in the system and keep your nose clean at
the same time."

Doyne had done well enough at Stanford to get into several graduate schools in physics. But one spin on his motorcycle around the
University of California campus at Santa Cruz convinced him to
move fifty miles down the coast to the shores of Monterey Bay. "I
was planning to be an astrophysicist, although I was still having
doubts when I came to graduate school. Did I want to be a physicist or not? I wasn't at all convinced it was the right thing for me
to do." He again faced a long grind through courses in differential
geometry, general relativity, and other subjects pertaining to the
motion of bodies inter- and extragalactic.

"I'm the kind of person who can sit and think about a problem
for twenty-four hours, until finally a switch goes off in my mind
and I get it. Or sometimes I dream the answer. But I totally panic
when someone puts a sheet of paper in front of me and gives me
an hour to get the solution written down." Doyne's first hurdle in
getting a doctorate in physics was a day-long battery of exams. He
spent a year preparing for them by learning how to pace himself
through mock tests. At the end of his second year at Santa Cruz,
he was the only member in a class of six to pass the exams.

"At that point, I was feeling really on top of it, good about physics and good about myself. I was accepted by George Blumenthal
as his student, and was headed down the road to becoming an astrophysicist like him, when I took off that summer for a Forest
Service job in Libby, Montana. That proved to be the start of my
rambling and gambling."

Norman Packard, meanwhile, had also launched himself from Silver City out into the great world. Declining scholarships from Cal
Tech and Stanford, he opted instead for Reed College in Portland,

Oregon. "I was attracted," he said, "by the thought of becoming a Renaissance person." The smaller community at Reed seemed more welcoming to a pianist-physicist and singer of Gregorian chants. Already an accomplished musician, Norman found Bach's Goldberg Variations the perfect antidote to occasional bouts of depression. He also took on the project, while studying Eastern philosophy, of developing a Zen self absent of ego.

The summer before leaving for Reed, on his way back to Silver City from a Rolling Stones concert in Albuquerque, Norman had driven off the road in a near-fatal car accident. He emerged from the hospital six weeks later with a piece of metal implanted in his leg and the emaciated look of a prodigy run amuck. He weighed 130 pounds and was still limping on a cane when he arrived for his first year of college.

"As I hobbled around campus, I immediately started a lot of rumors about who I was. I had skipped the first year of physics, and there were reports afoot about my being a genius. A skinny cripple doesn't have the easiest time breaking into college, but Reed has a high tolerance for weirdos in general. So I wasn't that out of place."

At the end of his third year in college — expecting like Doyne to make his fortune in the card parlors of the West — Norman also embarked on a summer of rambling and gambling.

2

Rambling and Gambling

No one can possibly win at roulette unless he steals money from the table while the croupier isn't looking.

Albert Einstein

In August 1975, at the Oxford Card Room in Missoula, Montana, Doyne Farmer introduced himself to the world of gambling. Nominally working that summer as a building inspector for the U.S. Forest Service — an assignment for which he had neither expertise nor reason to call on any — he had sat down to read A. H. Morehead's *Complete Guide to Winning Poker*. After memorizing the book from cover to cover, he emerged from his bivouac on Lake Koocanusa an accomplished poker sharp. His only problem lay in the fact that having never before played the game, his grasp of its actual mechanics was a bit weak. He fumbled his cards and dealt the deck wrong way around the table, but in five nights of gambling at the Oxford, he cleared more in poker winnings than his entire summer's salary from the Forest Service.

Sitting next to Doyne on his tour of the Montana card rooms was his friend Dan Browne, former motorcycle smuggler and Idaho Muscovite, who had originally lent him the copy of Morehead and convinced him there was money to be made in rambling and gambling. Browne is a natural card room talker. Give him silence and he'll fill it up. A chess player, clarinetist, and undergraduate physics major at the University of Idaho, he worked his way through college by commuting on weekends to poker games in Spokane, where he cleaned out the house in time to make it back to Moscow for an eleven o'clock class Monday morning.

Before walking into the Oxford, Browne and his protégé agreed on a strategy. "You can't tell someone you're a physics student," Browne said. "They just won't play with you." Sporting a vest and wool cap, he reassumed the childhood pseudonym of George "Bug" Browne, Bug for short. Complete with straw hat, suspenders, and southwestern twang, Doyne rechristened himself "Clem." Strolling into the Oxford, they sat on either side of a woman in a low-cut black dress. She tittered when Browne told her his name was Bug and then burst out laughing when Doyne introduced himself as Clem. He looked her straight in the eye and said, "Whatsa matter, lady? You don't like something about my name?" She blushed and quickly made her apologies.

"New Mexico Clem then settled down to some serious gambling," said Browne. "To look at him, you would've taken him for a veritable mark. He fumbled the cards and spilled his chips. He didn't know how to shuffle, and he dealt like someone just learning how to count. *One* and *two* and *three* and *four*. To the chagrin of everyone present but me, Clem turned out to be a pretty sharp player. After all, he knew every word that Morehead had ever written about poker, and he put it to good use."

After their nights at the Oxford, Clem and Bug drove to Lolo Hot Springs for a dip, some transcendental meditation, and a look at the sun coming up over the Rockies. "Everything grew timeless," Browne recalled, "as all we did was eat, sleep, and play poker." But a near disaster brought their tour to a sudden end.

Although Clem played a good game of poker, he had yet to acquire Bug's subtlety in disguising his winnings. "When I was starting out as a poker player," said Browne, "I used to bet only when the odds were on my side. But when you're taking money out of other people's pockets, they can get a little upset. I got thrown out of a card room once for playing too tight, which was actually a good thing for my game. It made me loosen up, get into the bantering and joking, gamble once in a while when the odds were even, throw a card against the wall and call it, buy everyone a round of drinks. When I started playing in Spokane after that I became one of the regulars, a good old boy who they would never think of throwing out of the room."

His last night at the Oxford, not even a monosyllabic drawl had escaped from New Mexico Clem as he sat on his cards for two hours straight. The hand he finally played was a wheel, 1-2-3-4-5, which is the best you can do in low poker. He raked in a sizable pot and headed straight for the bank. There was grumbling around the table about "card sharps" and "mechanics" when an ex–professional wrestler by the name of Emo leaned over and picked up the nine of clubs from underneath Doyne's chair. "I have no idea where it came from," said Browne, "but it took a lot of explaining to make that card look less than mortal. It was time to head back to school anyway."

Norman Packard had also lit out for a summer of gambling, and he too was playing by the book — Edward Thorp's *Beat the Dealer,* which outlines a card-counting system that Norman and a partner were implementing down at the blackjack tables in Las Vegas. After discussing the project with Ingerson and calculating the probabilities, Norman expected in two months' play to clear a cool ten thousand dollars.

The power of Thorp's card-counting system lies in its comprehensiveness, which Thorp had obtained through extensive use of computer simulations. While teaching at MIT, he had programmed a mainframe IBM 704 to calculate the shifting probabilities of winning at blackjack as the deck is dealt down from top to bottom. "It would have taken roughly ten thousand man-years to do the same calculations with the aid of a desk calculator," wrote Thorp of a machine that did the job for him in three hours.

Thorp's system relies on the fact that a player's chances of winning at blackjack vary according to the cards already dealt in previous hands. The house will win more often, for instance, if the aces have been exhausted from the deck. Allowing players to take advantage of these shifting probabilities, Thorp devised a "point-count system" and "basic strategy" of optimal responses. With perfect recall and faultless play, Thorp's card-counting strategy "is enough to give the player a comfortable 3 per cent edge!"

Thorp promoted his book in the early 1960s with much hoopla. Using capital supplied by professional gamblers, he had sat down at the blackjack tables in Las Vegas with reporters from *Time* and *Life* at either elbow. "A Prof Beats the Gamblers" announced an article in *The Atlantic Monthly*, although *Scientific American* was less sanguine in its appraisal of Thorp's achievement. "The system will benefit only the idle rich," it declared. "The edge it gives the player is so small that he needs a large initial capital and a lot of time to run up substantial winnings." Apparently, there were enough "idle rich" in the world to turn Thorp's book into a best seller and make the casinos take precautions against card counters. The owners introduced play with multiple decks, reshuffled cards after every deal, and forcibly ejected anyone suspected of card counting — Thorp included.

Norman knew that implementing Thorp's system in Las Vegas would be "a nontrivial problem." He discussed it with Len Zane, Ingerson's brother-in-law and a successful card counter, who also happened to be chairman of the physics department at the University of Nevada in Las Vegas. Zane passed on to Norman an enhanced card-counting system, written by the pseudonymous Lawrence Revere, which was specially tailored for playability in the casinos. "How to interact with pit bosses, how to win without antagonizing the management, these," advised Zane, "are crucial elements in any scheme." He helped Norman develop a costume that kept his bearded face masked by sunglasses and a straw hat with pink hatband.

As system players, neither Norman nor Doyne thought of himself as a gambler. They were scientists taking advantages of stochastic fluctuations. A gambling system, properly conceived, is anti-entropic. It locates fluctuating probabilities — shifts in the advantages or disadvantages of a game in play — and stores them up in small but consistent winnings.

"When you're operating a system," said Norman, "you can't be gambling. Any hint that you are means you're not playing the system. That's why to gamblers this approach is boring, pointless, stupid, and takes the fun out of the game. The thrill for them comes in being lucky, while for a system player there is no stroking your rabbit's foot and getting a big kick when you win. Your behavior is completely predetermined. You should play like an automaton."

Norman and his partner in Las Vegas, a classmate from Reed named Jack Biles, graphed their daily progress in the casinos. "There were amazing fluctuations," Norman discovered. "For five days straight the winnings graph would go up. We would shift to high stakes, and then the graph would go up even faster! But then invariably it sank in a slow but steady decline."

Near the end of the summer they visited the Gambler's Book

Club and "bought everything we could on playing cards in Las Vegas. We gradually came to the conclusion that we were being cheated. You can spot cheating by characteristic maneuvers on the dealer's part, and when we knew what to look for we found all the telltale signs.

"In our final tally we barely broke even. This means that our system was working to some extent, because a typical blackjack player goes to the tables and loses at a twelve percent rate, which is huge. We weren't losing that fast, but we weren't winning either."

Despondent over their bad fortune, Norman and Jack played their last hand and stopped for a drink at one of the casinos on the Strip. They sat at the bar and talked about the possibility of beating games other than blackjack. A bespectacled Texan with gold-rimmed glasses, Biles blinks and stutters with excitement over new ideas. Promiscuous with his suggestions, loving and leaving them without a thought, he declared, "I bet you can beat roulette using physics." There was a long pause during which Norman stroked his beard. "You're right," he finally said. "Use Newton's laws and find a way to enter initial data. It's classic physics."

"In the back of his book," Norman remembered, "Thorp announces in a cryptic sentence or two that he devised, but failed to implement, a system to beat roulette. I thought on reading this that it was utter hogwash. Roulette is a random game. You can't devise any betting scheme that will win. But on rereading Thorp, we realized that you might be able to develop a *predictive* scheme, and that *that's* what he was talking about."

Setting out to verify their hunch that roulette is a predictable game, Jack and Norman borrowed a tape recorder and stuffed it in a plastic bag. Standing next to roulette wheels in various casinos, they tapped the microphone on the tape recorder at each revolution of the ball in front of a fixed point. After transcribing these clicks onto graph paper, they found that the ball did indeed decelerate, fall off, and land in a regular way. Excited by their discovery, they thought with hindsight that blackjack and poker playing were naive propositions. Roulette was clearly the game to beat.

"I was supposed to rendezvous with Norman at the end of the summer and compare notes on our gambling experiences," said Doyne. "When I met him at the Greyhound bus station in Portland he was still wearing his straw hat and card counter's outfit. Rather than being despondent about his blackjack losses, all he could talk about was beating roulette. I told him it was a waste of time but he was so persistent he finally convinced me to think about it myself.

"I imagined a person eyeballing the wheel and making a click as the ball passed a fixed point. I assumed a tenth of a second as a fair guess of human accuracy. Then I made some back-of-the-envelope calculations to see what error this gives in predicting where the ball will fall off. The angle of error spreads out over time, but much to my surprise, the calculations looked good. It was theoretically possible to predict the final position of the ball within a few numbers. This realization was what first got me interested in roulette."

Norman, Jack, and Doyne spent three days working on the problem. How were they going to clock the ball and input data? Would they keep their hands in their pockets on a switch, or use their toes? Could they get a laser to throw an infrared beam across the path of the ball? Would ultrasonic sound follow it with a kind of Doppler effect? "We tossed out a lot of schemes," Doyne recalled, "and tentatively decided to go ahead with the project."

In the early division of labor, Jack and Norman undertook to build an electronic clock that would analyze their tape-recorded data. Back in Santa Cruz, Doyne and Dan Browne would study the project's feasibility. Their initial research centered on the problems of *scatter* and *bounce*. Scatter results from the fact that roulette balls, after dropping from orbit, sometimes find their trajectory interrupted by metal diamonds attached to the sloping side, or *stator*, of the roulette wheel. Bounce refers to the problem of balls hopping from cup to cup on the central disk, or *rotor*, before finally coming to rest. Bounce and scatter tend to randomize the game, and either one, if exaggerated, could put a wrinkle in Doyne's otherwise encouraging calculations.

To look more closely at wheels in play, Doyne and Dan Browne drove from Santa Cruz to South Lake Tahoe. "On our maiden trip to the casinos," remembered Browne, "our main goals were, one, to find out exactly what went on at the roulette table in a casino and, two, to make some observations — rough data taking, if you will, on the dispersion of the ball. How many cups did it tend to jump over before coming to rest? So with our notepads in hand, we stole from casino to casino, watching roulette wheels and taking down data. Doyne recorded exactly where the ball struck the numbers, and I noted which cup it actually landed in. Then we switched roles and repeated the process, until finally one or both of us had seen so many spinning wheels, red and black numbers, and little white balls, not to mention keno girls, that we couldn't see straight."

"Being paranoid about collecting data in the casinos," Doyne said, "we would remember a string of ten numbers and then duck into the toilet to write them down. Later we realized that lots of people were standing around with notebooks calculating mathematical systems, and that we looked even more suspicious running in and out of the toilet."

When they graphed their findings back in Santa Cruz, they discovered the possibility of gaining a clear advantage over the house. The ball is randomized by scatter and bounce, but not so randomized that predictability is eliminated. Many balls pass uninterrupted through the diamonds on the stator. Others land in a cup with nary a hesitation. Even the balls whose trajectory is altered are still fairly predictable.

"Though crude," said Browne, "these first data were very important. They showed that the ball bounced on average no more than a quarter or a third of the way around the wheel. If we could build a machine that could predict when a spinning ball would leave a circular track — a straightforward, although difficult, physics problem — we could beat roulette. Mind you, we were miles away from actually coming up with such a machine, but we knew it existed, at least in the realm of possibility."

They made another trip to Nevada the following spring. "Doyne and I took off in my Opel station wagon and headed for Paul's Gaming Devices in Reno. We had heard that Paul built professional roulette wheels and refinished old ones, and that his wheels were the standard models used in Reno and Las Vegas. We left Santa Cruz early, got to Reno around noon, and went straight to Paul's workshop. We gave him a phony line about needing a roulette wheel for a local fraternity party. I doubt he was fooled, although I'm certain he had no inkling what the wheel was really for. He showed us everything he had, and then we pointed to the best of the rebuilt models and said, 'We'll take that one.'"

Built in Detroit, inlaid with teak, ebony, mahogany, and other exotic woods, it was a B. C. Wills regulation roulette wheel. For the wheel, a top-of-the-line model newly refinished by Paul, and a wooden shipping crate, they paid fifteen hundred dollars in cash. "He thought we were odd," Doyne recalled, "because all we cared about was the condition of the track."

"Paul gave us a long spiel," said Browne, "about what a first-class wheel we had bought, about its heritage, precision tooling, and twelve kinds of African wood — although I still think for that kind of bread it should have come with four tires and a steering wheel."

Back in Santa Cruz they crated up the wheel and shipped it to Portland, Oregon, where Jack and Norman were going to study it with their electronic clock. The wheel got as far as Oakland when the trucking company phoned to say there was a problem with the delivery. On driving up to the Bay Area, Doyne was met at the loading dock by two FBI agents; they said the shipping of gambling devices across state lines was a matter of interest to them. Doyne protested that a roulette wheel like this, inlaid with twelve kinds of African wood, was nothing more than a collector's item, and convinced them to let him take it back to Santa Cruz.

Having passed his qualifying exams the previous spring, Doyne was supposed to be researching a dissertation in astrophysics. "I was trying to figure out the question of galaxy formation in the Hoyle-Narlikar cosmology. You start with a universe and it goes blooey. You then have matter homogeneously distributed throughout, which you want to clump into galaxies. But when you calculate how fast things should clump, they don't clump fast enough. This is the big test of a cosmology, to see if it allows for galaxy formation without having to resort to something hokey, like primordial black holes."

The study of physics divides into various intellectual and temperamental domains. "If you want to be on the philosophical forefront of physics," said Doyne, "you go into particle physics. It's the esoteric branch that is also intensely cutthroat and deeply established. It is not practical. The practical people go into solid-state physics." The least practical and most romantic of all physical domains is cosmology. Doyne later in his career would have the chance to shift from romance to revolution. He and several colleagues at Santa Cruz would find themselves at the center of what Thomas Kuhn called a paradigm shift: a radical shakeup of the very categories used to think about physics. At the moment, though, not even romance was going well. Doyne was bored with school and distracted. Far more interesting to him than Hoyle-Narlikar was the question of galaxy formation in the roulette cosmos.

As a preceptor at Cowell College, one of the affiliated schools scattered among the redwoods at UC Santa Cruz, Doyne lived in a two-room apartment attached to a dormitory. "A big problem with doing experiments in Doyne's rooms," said Browne, "was how to keep the wheel hidden. We didn't want anyone finding out what we were up to. But with people wandering in and out, it was hard to disguise an ashtray three and a half feet in diameter."

The problem was further complicated when Doyne's mother came to visit for a week from Fort Smith, Arkansas, where she and her husband had moved to open a pie shop meant to capitalize on her secret recipes and his innovations in culinary engineering. As one good talker appreciating another, Dan Browne characterized Mrs. Farmer as "a nice, friendly woman, crazy as a loon, sharp as a tack, who talks a *blue* streak, way up into the miniwavelengths; so we weren't sure it was the right time to clue her in to the project, especially one as bizarre as this."

Each night Doyne and Browne moved the wheel over to the experimental physics laboratory, which Doyne had access to as head teaching assistant responsible for setting up daily experiments. Here they worked from eight at night until two or three in the morning. Using Norman's electronic clock, as well as infrared detectors and high-speed film, they measured the physical forces at work in roulette. The ball's trajectory, deceleration, scatter, bounce, and relative position vis-à-vis the spinning rotor were all quantified into the numerical stuff of physical prediction.

"I have fond memories of those late nights in the physics laboratory," said Browne. "Afterwards we went skinny-dipping in the university pool, and then made it down to Ferrell's doughnut shop just as the cinnamon rolls were coming out of the oven. We had only one close call after we started carrying our crate around campus. A janitor by the name of Fred Faria, who called Doyne 'Mr.

Professor,' was dying to know what we did every night in the physics laboratory. He finally caught us with the wheel uncovered. He was impressed, but not overly taken aback. He thought we were running a gambling ring to pay our way through school. No big deal."

As his academic progress reports slipped from "excellent" to "satisfactory," Doyne finally came clean with his adviser. "I told him what was going on, that I wasn't interested in doing astrophysics but wanted to take a year off to do roulette. I explained that that's why I had been screwing around the past four months, because I had been spending all my time working on roulette."

George Blumenthal, Doyne's adviser, was sympathetic. He reviewed the equations and thought the scheme looked promising. He then told Doyne about his own career as a blackjack counter, which had ended abruptly on his being wiped out by a cheating dealer.

Late that spring there gathered in Doyne's apartment a nucleus of people prepared to work throughout the summer building a predictive machine to beat roulette. Norman Packard and Jack Biles came down from Portland to join Dan Browne, who was already camped in Santa Cruz on Doyne's floor. Another physics student, John Boyd, arrived from Moscow, Idaho. The lone humanist among them was Steve Lawton, a friend of Doyne and Letty's from their undergraduate days at Stanford.

Tall, balding, gregarious, and more athletic than his spacy demeanor might otherwise have indicated, Lawton was a specialist in utopian literature. That spring he organized a reading group on the subject, and when not acting as gofer on shopping trips to the Silicon Valley, he led discussions on how best to organize utopian communities. This mix of theory and practice characterized life in Doyne's apartment during the early stages of what was codenamed Project Rosetta Stone, or the Project, for short.

"I used to think a lot about community," said Doyne. "How to bring people together and make things happen. We wanted to build up a network of people we could trust and a set of tools. We started the Project thinking it would be a way to organize this sort of community. It would finance a home base for everyone. I may not touch home very often, but there has to be someplace to go back to. Otherwise it's crazy in this society, where people drift off to Timbuktu and Philadelphia just because that's where they get jobs. There has to be a better way for people to control where they live and how they stay in touch with each other."

"We imagined the roulette project as a cash cow that would support our other interests," said Norman. "Ever since our days in Silver City we had had the idea of starting a company to finance our electronics projects and travel to faraway lands and the study of physics. It's been a long-standing pipe dream to put together a community that would allow us to live off the fruits of our ideas."

Tom Ingerson, then on sabbatical at the Cerro Tololo Observatory in La Serena, Chile, was busy tracking astral bodies known as Seyferts across the night skies of the Southern Hemisphere. His spirit was often invoked in these early discussions; his practical advice took longer to arrive by mail. In dense epistles that filled both sides of five or more typewritten pages, he melded philosophy with physics and made numerous practical suggestions that were later adopted by the Projectors.

Their immediate tasks ranged from basic research to the design and construction of a computer. They needed investors, lines of credit, a checking account, stationery. Who would direct the Project, and how, eventually, would they reward all the people contributing either time or money? They began by taking out papers of incorporation. In spite of its official status, this company was to be run democratically, with all decisions arrived at by consensus and all future profits divided equitably between investors and workers.

"We needed a front for dealing with the outside world so that we could do things, like buy electronics," said Norman. "We had to come up with a name and get a checking account under that name and keep track of all the money involved, because we were trying to be conscientious about money. We had in front of us the vivid image of profits. We wanted to keep the record straight about who would be getting what, so there wouldn't be any conflict later on when the Project was earning hundreds of thousands of dollars."

Doyne was leafing through the dictionary one day when he found the word *eudaemonia*. Aristotle posits the existence of many daemons, or presiding spirits, but his favorite among them was the *eu*daemon, or spirit of rationality. Eudaemonia describes that special happiness resulting from an active, rational life. Doyne's dictionary defined it as "a state of felicity or bliss obtained by a life lived in accordance with reason." When put up for a vote against Utopian Ventures and Amphibian Productions, Eudaemonic Enterprises (the first word is pronounced something like a Wagnerian recitative, *yoo-die-mahn-ick*) was chosen unanimously as the most suitable name for the new company.

Having at this stage of its existence no capital, retained earnings, or other financial assets, the best Eudaemonic Enterprises could offer its worker-owners in terms of payment was a share of future profits. This share would be sliced out of something called the Eudaemonic Pie. Filled with the totality of all future earnings, the Pie would be served out in portions calibrated to the number of hours or amount of money invested in the Project. All tasks, from building microcircuits to spinning roulette balls, would be rewarded equally. Outside advice on designing equipment, technical reviews of the physics involved, suggestions on how to implement the Project — these and other ideas would also be assigned hourly values and given a piece of the Pie. No one doubted that this conceptual dish would someday be rich enough to feed everyone who helped to make it.

Eudaemonic Enterprises imagined itself ultimately developing a number of pies. Project Rosetta Stone would provide the capital for launching other ventures, which ranged from the design of dirigibles and weightless cubes to the building of energy-efficient houses and a utopian lending library. Money from roulette would also go toward buying land in the Coast Range of Washington or California. Here, E.E. would construct its own utopia: a technological commune of friends and tools gathered for the purpose of putting science to human use. Between visits to Timbuktu and Philadelphia, all Eudaemons would have someplace to call home.

"The assembling of all the Projectors was an ecstatic occasion," said Norman. "We were electrified and ready to blast off on this promising endeavor that wasn't just another kinky way to spend our summer, but potentially a gateway to new levels of life."

While speculating on the fruits of Eudaemonia, the Projectors more immediately set to work studying the game of roulette. "We began with a thorough feasibility study," said Doyne. "We had to figure out the trajectory of the ball. Then we needed to pin down the problems of bounce and scatter. At the same time we had to research the hardware necessary to implement the system. What kind of computer would we build to input data and output predic-

tions?" There was the ultimate problem of the equations themselves. What algorithm would integrate all the forces in the game
with enough accuracy to predict its outcome?

When not working nights in the physics laboratory, the Projectors slept on Doyne's floor or out under the redwoods. Tom Ingerson had given Doyne the Blue Bus as a graduation present, and
Norman now moved into it with Lorna Lyons, a friend who had
come to visit from Portland. Statuesque and imposing, Lorna has
a face that looks from certain angles like a Leonardo portrait. Her
voice is more American in scope. Imagine a Chicago grain report
on sexism or mind-altering drugs. While shuttling around campus
between the university's long- and short-term parking lots, Lorna
and Norman inaugurated what would become a long-lasting affair.

Later in the summer, when Professor Nauenberg in the physics
department rented Doyne his house on Laurent Street, the Projectors moved to more spacious quarters overlooking the town and
Monterey Bay. They divided tasks among themselves as follows:
Dan Browne, Steve Lawton, and John Boyd would undertake physical studies of the ball's trajectory and bounce. Norman Packard
and Jack Biles would research computers and hardware. Doyne
Farmer would analyze data and derive the roulette equations. "We
expected to finish by the end of the summer," said Norman. "Certainly by the following Christmas we planned to have the whole
thing working and ready to take into the casinos." Beating roulette
proved more difficult than Norman imagined. His estimate was
overly optimistic by twelve months. It was not until the *following*
Christmas that they reached Nevada with their first computer.

Finely machined and oiled, the disk of a roulette wheel suffers little
decay in its velocity. A roulette ball, on the other hand, confronts
numerous opportunities for entropic degradation. As it drops from
orbit and arcs toward a rendezvous with the spinning rotor, it
passes through a veritable minefield of galactic booby traps. It encounters friction from the track on which it turns. It faces wind
resistance, drag, and the ineluctable pull of gravity. Analyzing the
ball's trajectory would still be relatively straightforward were it
not for further complications. Several metallic diamonds, pitched
either horizontally or vertically, decorate the stator of a roulette
wheel. On hitting a diamond, as the Projectors had already discov-

ered, a roulette ball will be knocked off course. It may be lofted higher in its trajectory, or dropped straight toward the wheel, where it faces even more interference from the metal frets that separate the cups.

"There are several questions involved in whether or not the game is predictable," Norman explained. "First you have to ask whether the wheel itself is predictable, which means that if you input two clicks into the computer, can it tell what the position of the wheel is going to be several revolutions hence? It's clear that if you input the clicks accurately enough the computer can do it, but then you have to find out whether the clicks *plus* the error we introduce by being a little spastic, because we're human beings, are *still* accurate enough for prediction.

"Then you have to ask the same question for the ball. Does the ball, as it's running around the track, slow down uniformly? If the track is rough or irregular, this may not be the case. The last and major question: If we can predict the rotor motion and ball motion so that we know exactly *when* and *where* the ball is going to fall off and exactly *where* it's going to hit the rotor — suppose we give ourselves that predictive power — then what happens? What happens is that the ball bounces around a lot. If it bounces around too much, then all of your predictive power up to that point is useless."

Set up in the basement and covered with electronic gadgetry, the B. C. Wills roulette wheel underwent a series of tests that left it looking like a coronary patient in intensive care. The experimental wing of the Project began by stationing photo resistors around the wheel. In conjunction with little light bulbs beamed onto the track, the resistors recorded the ball's rolling trajectory. "Running our experiments at night," said Doyne, "it looked like Halloween, but we still weren't getting sharp enough signals to trigger the timer on Norman's clock."

They then discovered optrons: tiny infrared-sensitive devices that work at close range like radar. These incorporated light-emitting diodes that sparkled ruby red as the ball passed in front of them. "We set these up at eight stations around the wheel," said Doyne. "Because of the jungle of wires feeding from the optrons into an amplifier and then into Norman's clock, each station was about as big as a cigarette pack. The optrons triggered whenever the ball passed in front of them. By noting the elapsed time of the ball from one station to the next, we could easily compute its velocity."

The Projectors measured the ball's scatter and bounce by means of a device they called the guillotine. This consisted of a scaffold built over the wheel on which were hung a camera and stroboscopic flash. The camera, the flash, and Norman's clock were triggered by a second amplifier that picked up the sound of the ball as it hit metal. These photographs recorded the first strike and subsequent landing of the ball on the rotor. By plotting on two axes this relationship between *strike* and *land*, one could graph the ball's average displacement, either backwards or forwards, along the wheel.

This graph of a frequency distribution is known in statistics as a histogram. Variances in behavior are plotted on the x axis, and their frequency of occurrence on the y axis. Other than a *strike-land* histogram, another variety of obvious interest to the Projectors — especially when they had finished building their computer — would record the ball's displacement either backwards or forwards from the predicted number. Over the life of the Project they would record hundreds of these *predicted-actual* histograms.

In the early strike-land histograms, where only the behavior of the ball was being studied, how far and in which direction it bounced were irrelevant. The Project cared only about consistency. "What we wanted to see in the data," said Norman, "was a sharp peak, a tendency on the part of the ball to displace itself with regularity in one direction or the other. Instead, we got something looking more like a hump. Predictability in roulette is inversely proportional to the width of that hump. If the hump spreads out over more than nineteen numbers, that is, over more than half the wheel, it gets very hard to predict where the ball will land."

The Projectors also examined the nature of roulette balls themselves. These are made out of substances ranging from ivory to human bone, although more common materials include nylon, Teflon, acetate, and the composite material of billiard balls. These substances vary greatly in their characteristics. Teflon decelerates 100 percent faster than the billiard ball composite, while nylon has more bounce than the others. The standard ball in Las Vegas is made of acetate, which was lucky for the experimental team. A good reflector of infrared light, acetate balls are easy to track.

"Along with gathering data, I was thinking a lot about the theory of roulette," said Doyne. "What kind of equations does the roulette ball satisfy? A rolling ball on a circular track is no problem, but in most physics classes you don't work much with friction. You seldom get beyond an incredibly simplistic model that takes account of either static friction or sliding friction. I had assumed the main friction would be between the ball and track, until it occurred to me to think about *wind* friction, which is proportional to the square of the velocity. I estimated what this coefficient of friction would be and discovered to my surprise that wind resistance alone is mainly responsible for the ball slowing down."

The Projectors next discovered the importance of *tilt*. An ideal roulette wheel spins on a perfectly level plain, although no such surface exists in the casinos of this world. Having set their wheel on stainless steel jacks, the Projectors had spent hours trying to level it into a semblance of perfection. "Then I did a little calculation," said Doyne. "Given x amount of tilt on the track, what difference does it make in the fall-off velocity of the ball? It turned out that a tenth of an inch was enough tilt to make a significant difference in where the ball came off the track. It's very hard to get the wheel level within a tenth of an inch. So it was obvious from that moment that tilt would also play an important role in the prediction."

Given these variances in tilt, bounce, scatter, and drag, the Projectors discovered further wrinkles in the roulette cosmos when they rented a second wheel from a novelty store in San Francisco. Rigged up under the guillotine for high-speed filming, the wheel revealed two further singularities. The San Francisco wheel was a wreck of its former self. A wobble in the central spindle slowed it down more quickly than the B. C. Wills wheel. The track was badly pitted, so that balls spinning on it would again slow down and fall off more quickly. This meant that the final algorithm for predicting roulette, if flexible enough to account for differences such as these, would require what are known in physics as *adjustable parameters*.

"When we rented the second wheel," said Doyne, "we discovered that the parameters for wheels on which we might play were going to be different. That meant I had to sit down and derive mathematical functions for these parameters which would be adjustable. And then I had to think of a way to interact with the computer program so that it could change these parameters every time we went up against a new wheel."

Doyne isolated no fewer than five variables in need of adjustment. There were two for the ball itself: one for measuring the rate at which it slowed down, and the second for establishing the mean velocity at which it fell off the track. He called the first of these the *ball deceleration parameter*. The second was known as the *time of fall parameter*. These shifted according to the kind of ball in play, or the curvature and condition of the track itself. Another variable measured the rate at which the cups on the central rotor slowed. Two final parameters dealt with tilt. They approximated its magnitude and swung the computer's prediction around to the high side of the wheel, which is where a roulette ball tends to come off the track.

While the roulette equations were being worked out in detail, another team, directed by Norman Packard and Jack Biles, researched their implementation. It was one thing to have a wheel sparkling with optrons, but something more subtle was required for playing roulette in Las Vegas. Attracted to the exotic in technologies, Jack suggested they use lasers or radar. A simpler solution, however, lay in microswitches operated by fingers or toes. Two clicks recording successive passes of the ball in front of a fixed reference point would establish its position, velocity, and rate of deceleration. "Assuming," as Norman put it, "that we weren't too spastic in our clicking."

The Projectors mounted a series of tests known as the Human vs. Machine experiments, which were conducted with an eye-toe coordination device that also functioned as a biofeedback machine for improving human reflexes. While clocking the ball around the track, humans tapping microswitches were pitted against optrons. The humans lost, of course. But their performance looked remediable. That the more athletic among them did better than the merely cerebral suggested that with proper training, humans using microswitches would be adequate to the task.

Without knowing what a roulette computer would look like, or the final content of its program, the Projectors decided in theory that the system should be operated by a two-person team. A data taker, standing near the wheel, would spend a few minutes adjusting the necessary parameters, and then clock the ball and rotor. An apparently unrelated bettor, standing far down the table but actually linked to the roulette computer by radio or some other connection, would receive the predictions and use them to play a high-stakes game.

While brainstorming output devices and various ways to link data taker and bettor, the Projectors wrote to hearing aid manufacturers and collected brochures on ultrasonic technology. They visited a hospital supply company for a demonstration of electroshock equipment. They considered polarized eyeglasses, laser detectors, light-emitting diodes built into wrist watches, and radio waves of every conceivable frequency. Finding the perfect link — reliable yet undetectable by the casinos — would require some ingenuity.

Working around the clock on guillotine and high-speed photos, optrons and histograms, the Project generated a mass of data on everything from wind resistance to human perception. As chief theoretician, Doyne was supposed to figure out what it all meant. His assignment was to isolate the equations of motion that govern the individual parts of roulette, and then integrate these into one synthetic equation, or algorithm, capable of prediction. It was at this point that Doyne, for the first time in his life, resorted to using a computer.

"He had made it a policy throughout his undergraduate career," said Norman, "never to use a computer. It was a matter of honor. Physicists disdain calculator-minded people, the class of which is epitomized by the engineer. This disdain is rooted in the fact that true understanding of the physics of a problem does not depend on the particular numbers involved. Doyne hadn't even used a calculator in doing his labs. He always preferred a pencil and paper."

Others in the group who were conversant, if not fluent, in its several languages, nudged Doyne toward the computer. Working at the university on a PDP 11/45 built by the Digital Equipment Corporation, he took a week to teach himself how to program in BASIC. Doyne and other physicists with whom he worked at Santa Cruz would later emerge as masters of the new technology, but even at this stage he was impressed by what computers could do. Simple, yet vital things: like simulate human error.

Eudaemonic research proceeded with the casual mania peculiar to this part of the world. Nude sunbathing on the back deck was combined with phone calls to Advanced Kinetics in Costa Mesa, American Laser Systems in Goleta, Automation Industries in Danbury, Connecticut, Arenberg Ultrasonics in Jamaica Plain, Massachusetts, and Hewlett-Packard in Sunnyvale, California, where Norman Packard's cousin, David, presided as chairman of the board. The trick was to make these phone calls at noon, in the hope that out-to-lunch executives would return them at their own expense. Eudaemonic Enterprises, for all they knew, might be a fast-growing computer company branching out of the Silicon Valley. Sniffing the possibility of high-volume sales, these executives little suspected that they were talking on the other end of the line to a naked physicist crazed over roulette.

By the end of the summer, Professor Nauenberg's house had been transformed from top to bottom into a working physics laboratory. Resident researchers could be found day and night huddled around their roulette wheel laden with a guillotine, stroboscopic flashes, cameras, optrons, microswitches, an electronic clock, and roulette balls of every size and composition. Another problem in applied physics — how to keep two particles from colliding — resulted in the house being divided into separate zones: one for Rembrandt, the professor's German shepherd, and the other for Pate, Jack Biles's thirteen-year-old beagle, who scored a decisive victory in the summer's running dog fight.

The dogs savaged each other and their respective domains. The lawn turned brown and died. The house fell into a state of chaos. "Pie, spaghetti, ice cream, anything," said Lawton, "would be consumed for breakfast. It was like living with a gaping void into which all food disappeared. There was that basic physicist's oblivion. They could eat a bowl of caviar or potato chips and an hour later not know the difference."

Living and working together, the Projectors got to know each other better than they might have wished. As the feasibility studies expanded to measure differences in bounce between nylon and Teflon roulette balls, Jack Biles grew impatient. He was in a rush to head for the casinos with a calculator in a paper bag and do his feasibility studies with a pile of chips in front of him. "Jack was content with making a brilliant point or two," said Lawton, "but not much help in putting it together. He preferred to talk about ideas rather than implement them."

"There were two schools of thought that summer," said Norman.

"The 'Be Careful School,' under whose auspices the tests were being performed, wanted to know whether prediction in roulette is theoretically possible before we went off and built a microcomputer. The other school said, 'We don't have time for that. You can tell from looking at the regularity of the ball that the game is predictable. Let's build a computer and measure our advantage from how well it does.'

"The problem with Jack's approach is that you wouldn't know in principle whether prediction is feasible. This means you could never distinguish between a computer that didn't work and a game that isn't predictable in the first place, and that sort of thing made us very uncomfortable."

Described by one observer as "jazzed all the time about new ideas, your classic boy inventor having fun with his off-the-wall enthusiasms," Jack became increasingly disenchanted. "He was feeling jilted by it," said Norman. "The Project had been taken out of his hands. He left for Oregon at the end of the summer, and the longer the Project dragged on in Santa Cruz, the farther removed he got from it."

If Biles was always in a rush, his antithesis in Santa Cruz was John Boyd, commonly known as Juano. Tall, weedy, with smudgy black glasses and lank hair falling into his face, "Juano was your basic seen-but-not-heard character," according to Lawton. "Once you got him started on a project, he would work limitlessly on it. He was the opposite of Jack. Once in motion, he would continue in motion, but when he came to a stop, he had to be recharged."

As for the two dynamos on the Project, Doyne and Norman, "We really buzzed," said Doyne. "We were working like hell and enjoying it thoroughly. Norman especially amazed me. I couldn't believe how anyone could work so many hours a day and be so congenial about it. That summer he even managed to begin his romance with Lorna. For me at least, Norman set the pace."

Where Norman as a worker was rock-steady and indefatigable, Doyne was manic. His mind yo-yoed from Newtonian mechanics down to the minutiae of ordering transistors from jobbers in Sunnyvale. "His organizing ability came to dominate," said Lawton. "He could organize five people at once when they were barely able to organize themselves. He had the ability to provide a kernel of initiative, while still leaving room for other people to act."

By the end of the summer, the feasibility studies into bounce, scatter, friction, wind resistance, tilt, and other parameters unique to individual roulette wheels showed conclusively that the game was predictable. Given a powerful algorithm capable of integrating these forces, Eudaemonic Enterprises confirmed that one could gain a whopping 44 percent advantage over the casinos. Now all that remained was to formulate the algorithm and build the predictive device itself.

Jack Biles suggested they strip down and reprogram an electronic calculator, although he agreed later that this would have been a kludge, a baling-wire approach to a problem that called for something more elegant. The Project opted instead for a relatively new and esoteric technology. They would devise a program for a microprocessor and build it into a computer small enough to operate in the casinos without detection. As far as they knew, their microprocessor would be the first to pioneer a trail from the Silicon Valley to the roulette tables of Glitter Gulch.

"Given the existent technology," said Norman, "the computer we built is the ultimate predictive machine. You could ask any state-of-the-art electronics engineer to design it, and this is what he'd come up with. Ours is the rare example of a task solved by exactly the optimum technology."

"By the end of the summer," said Doyne, "we had a vague idea of the components we needed to get the computer going. We counted up the chips, and it looked as if by designing our own system, without any superfluous chips for a keyboard or tape recorder or LED display, we could build a computer as small as a cigarette pack, which turned out to be true."

Except for the roulette wheel, everything required for the Project had been homemade. So too would their first computer. From a mail-order catalogue Eudaemonic Enterprises ordered a microprocessor chip and a development kit that promised everything needed for building a bare-bones computer. The beauty of it lay in its flexibility. This chip could be programmed to do anything. The horror of it lay in its ignorance. Arriving with no program at all, even the multiplication of one times one would be unknown to it. Before engaging it in the higher math of roulette, someone would have to teach this computer how to do arithmetic.

3

Driving Around the Mode Map

Berlin is a nice town and there were many opportunities
for a student to spend his time in an agreeable manner,
for instance with the nice girls. But instead of that we
had to perform big and awful calculations.

Konrad Zuse

It is a curious fact that the inception of the computer, the game of roulette, and the basic laws of probability are all attributed to the seventeenth-century French mathematician and philosopher Blaise Pascal. Pascal was also the gambler who proposed in his famous "wager" on the existence of God that "One must bet on it." It was a less existential Pascal who in 1642, at the age of nineteen, invented the mechanical adding machine. His father, provincial administrator in Rouen, had drafted him into tallying the year's income tax. For Pascal junior the business was sheer drudgery, but out of this compost of necessity and boredom sprang his first great invention.

Pascal realized that numerical digits could be arranged on wheels in such a way that each wheel, in making a complete revolution, would turn its neighbor one-tenth of a revolution. While viewing the mechanism through a little window, one dialed the answer to problems in summation or subtraction. The genius in Pascal's invention — which remained the basic concept employed by mechanical calculators from his day to the present — lay in the transposition of arithmetic functions onto the physical locations of a machine. To get from there to the modern computer, one simply adds electricity and converts the cogs of a Pascaline into electronic charges stored in the crystalline structure of silicon.

Thirty years after its invention, Gottfried Wilhelm von Leibnitz made the first major improvement to the Pascaline by adding what came to be known as the Leibnitz wheel. This enabled the machine to do multiplication and division, as well as addition and subtraction. The next significant redesign was attempted in the nineteenth century by Charles Babbage, inspired inventor of the speedometer, the cowcatcher, and the first reliable life expectancy tables. Babbage spent the last thirty-seven years of his life, until his death in 1871, forging the cogs and rods of a great Analytical Engine. The machine possessed, in its essential design, all the features of a modern computer. These consisted of a logic center, which Babbage called the "mill," a memory, known as the "store," a control unit for carrying out instructions, and a system of punched cards similar to those used in Joseph-Marie Jacquard's looms for inputting data. But it was not for the age of steam power and mechanical gears to realize a machine as complex as this, and Babbage died with most of his project unfulfilled.

At this stage in its history the computer joins its destiny to the fortunes of war. Born to Athena — patron of spinning, weaving, cities, and bureaucracies — the computer grows up the stepchild of Mars. Babbage's dream materialized only to defend national interests on either side of the Maginot Line. In 1936 the young engineer Konrad Zuse filled his parents' Berlin apartment with a computer built out of scrap parts and a German Erector Set. Successors to Zuse's Z1 computer, constructed more reliably with electromagnetic telephone relays, were used by the Nazi war machine for aircraft and missile design, where their prowess in number crunching brought them to the attention of Adolph Hitler. Advised to mount a crash program for building more of Zuse's computers, the Führer made a tactical — and, for us, fortunate — error in thinking he could win the war without them.

In the meantime, the British early in the war had gathered an ace team of mathematicians and chess players at a country house in Hertfordshire known as Bletchley Park. Their assignment was to crack the German codes generated by the so-called Enigma Machine, one of which had been captured and sent to England by the Polish secret service. Only a computer could unscramble codes as complex as the Enigma's, and the Bletchley Park crew succeeded admirably in building a number of crude but effective decoding machines. These included the Colossus and its ten successors, which were the first computers to use vacuum tubes rather than switches or relays for shuttling the on-off charges by means of which modern computers think. Unlike Pascal, Leibnitz, and Babbage, both Zuse and the Bletchley Parkers — with Alan Turing the most notable among them — had substituted base two for base ten as the principal counting unit in their computers. This allowed for a dramatic leap in the speed with which information could be processed. Flashed from tube to tube in series of 1's and 0's, the stuff of missile design and code breaking could now be pulsed through electronic circuitry at the rate of two hundred logical decisions per second.

Along with Zuse and the British, the Americans also understood the wider application of computers to war. Howard Aiken, on leave from the navy, finished building the Mark I at Harvard in 1943. This was an electromechanical machine in which "the gentle clicking of the relays" sounded to one observer "like a roomful of ladies knitting." Intended for the computation of ballistics tables, Aiken's computer was quickly outmoded by a far more efficient machine built with vacuum tubes at the University of Pennsylvania. The ENIAC, or Electronic Numerical Integrator and Calculator, weighing thirty tons and holding eighteen thousand tubes, spent its early life crunching out gunnery tables for the Aberdeen Proving Ground in Maryland.

In describing the advent of the computer as a "revolution," we tend to forget what it initially revolutionized. As Joseph Weizenbaum, professor of computer science at MIT, put it: "The computer in its modern form was born from the womb of the military. As with so much other modern technology of the same parentage, almost every technological advance in the computer field, including those motivated by the demands of the military, has had its residual payoff — fallout — in the civilian sector. Still, computers were first constructed in order to enable efficient calculations of how most precisely and effectively to drop artillery shells in order to kill people. It is probably a fair guess, although no one could possibly know, that a very considerable fraction of computers devoted to a single purpose today are still those dedicated to cheaper, more nearly certain ways to kill ever larger numbers of human beings."

Computers were held in bureaucratic captivity from the 1940s to the 1970s, but somewhere along the line — one can date their ultimate escape from the invention of the microprocessor — they broke free of martial law and opened themselves up to the zaniness and democratic efflorescence whose truly revolutionary applications we are only now beginning to see. The microprocessor, which is nothing more than a chip of silicon etched with the geometries of memory, gave the slip to the authorities and their central processing units. This new technology effected a fundamental shift away from mainframe, centralized, stationary computers protected by hierarchies of protocol toward bite-sized, transportable, independent, and democratic computers capable of functioning entirely on their own. With the advent of the microprocessor in 1970, anyone, at least in theory, could walk around with the power of an ENIAC snuggled into a shoe. Once liberated for the work of eros and free play, the computer could develop a talent for games, poetry, music, and — as befits its Pascalian origin — the playing of roulette.

The prerequisite for building an ENIAC in a shoe was the computer's miniaturization. This was another technological shift spun off from the military, particularly its space wing. As they went in hot pursuit of Sputnik and other galactic menaces, NASA and the air force needed computers light enough for liftoff. Prime contractors for the military obliged by reducing the size of their product with three remarkable advances in as many decades, and it was the third of these advances that allowed the computer to make its final, definitive break for freedom.

In the 1950s the transistor replaced the vacuum tube. Shuttled among junctions located in the structure of silicon crystals, current pulsed through a *trans*fer re*sistor*, or transistor, could amplify sound or switch signals in a hundredth the space of the old tube technology. The second breakthrough came with large-scale integration (LSI), a new technique that allowed for circuits made up of *thousands* of transistors to be etched onto wafers of silicon the size of a fingernail. The final, liberating stroke came when Ted Hoff, an engineer at Intel Corporation in Santa Clara, California, figured out how to fit *all* the math and logic circuitry of a computer onto a *single* chip of silicon. "A true revolution," is how Robert Noyce, co-inventor of the integrated circuit and a founder of Intel, described this culminating event. "A qualitative change in technology, the integrated microelectronic circuit has given rise to a qualitative change in human capabilities."

Hoff was thirty-three years old when he invented the microprocessor. Not long out of Stanford — one among hundreds of bright students graduated from the physics department into the high tech factories along the Camino Real — he was eager, as he said, "to get out into the commercial world and see if my ideas maybe didn't have some commercial value." Noyce had turned him loose on a problem raised by some Japanese manufacturers of desk-top calculators. They wanted a calculator with math and logic circuits etched onto no more than eleven chips. It was while lost in the maze of designing this circuitry that Hoff came up with a new way of thinking about the geometries of silicon itself.

By adding extra dimensions to an already infinitesimal universe, he saw how to condense the eleven chips into *one* microelectronic circuit that would constitute the central processing unit, or CPU, of a new kind of computer. This "computer on a chip," as Intel called it, required the support of additional chips providing it with memory and a program, as well as input-output circuits and a clock to synchronize its operation. But Hoff's microprocessor — built on a single chip of silicon measuring one-eighth by one-sixth of an inch, and holding no fewer than 2250 microminiaturized transistors — sprang to life as the fully developed mind of a working computer.

At first not even the military knew what to do with the microprocessor. They had yet to comprehend the truth in Noyce's prediction that "the future is obviously in *decentralizing* computer power."

Intel, which previously had specialized in making semiconductor memories for computers, quickly became the world's largest producer of computers themselves. With no idea as to who might buy such an exotic device, they christened their first microprocessor the 4004, wired it onto a plastic board the size of a paperback book, attached a clock, control devices, and four more memory chips, and launched the MCS-4 out onto the market as the world's first "microprogrammable computer on a chip."

Its beauty lay in its flexibility. Teach a microprocessor how to add, instruct it in the solving of differential equations, program it with an algorithm capable of integrating Newton's laws of motion, and it can land a spaceship on Mars, or play roulette. "An individual integrated circuit on a chip perhaps a quarter of an inch square," wrote Noyce, "can now embrace more electronic elements than the most complex piece of electronic equipment that could be built in 1950. Today's microcomputer, at a cost of perhaps $300, has more computing capacity than the first large electronic computer, ENIAC. It is twenty times faster, has a larger memory, is thousands of times more reliable, consumes the power of a light bulb rather than that of a locomotive, occupies $\frac{1}{30,000}$ the volume, and costs $\frac{1}{10,000}$ as much. It is available by mail order or at your local hobby shop."

On September 8, 1976, in a package addressed to Mr. "Dwang" Farmer, Eudaemonic Enterprises received in the mail its first computer. Shipped from MOS Technology, Inc., in Norristown, Pennsylvania, the package contained a KIM, or *K*eyboard *I*nput *M*odule, computer development kit. For the price of two hundred fifty dollars, the kit included an Intel 6502 microprocessor — the same one later used to make Apple computers — a second chip to serve as a memory, two more chips specialized for inputting and outputting data, a crystal clock, an interface that allowed the computer program to be stored in a tape recorder, a primitive Korean keypad, a panel with enough light-emitting diodes to display one line of numbers, and a plastic board on which to solder all these parts together. With the addition of four extra memory chips, Eudaemonic Enterprises had purchased, for a total of four hundred dollars, a computer intelligent enough to navigate the Newtonian cosmos.

Taking over the work done previously by the university's PDP 11/45, the KIM would serve as the Project's mother computer. The reason for switching from mainframe to micro lay in the latter's adaptability. For another few hundred dollars, the Project could build onto the KIM a device for *burning* — or programming — secondary chips. Capable of reproducing itself ad infinitum, the KIM, once successfully programmed with a roulette algorithm, could be cloned into the smaller computers that would actually be carried into the casinos. The KIM was going to be the Ms. Big of the operation, masterminding the work of her minion computers in Las Vegas while cooling off under the redwood trees in Santa Cruz.

When soldered onto its plastic card, the KIM functioned as an eight-bit microcomputer with five kilobytes of random-access memory. In learning the language of bits and bytes, remember that digital computers operate in the rudimentary world of base two mathematics. They manipulate *bi*nary dig*its*, or *bits* of information. They think in spindly strings of 1's and 0's, which are themselves the symbolic representation of electrons passing through transistors. Thousands of transistorized locations are etched into a minute piece of silicon. Each of these locations in turn can be oriented as either "on" or "off," 1 or 0. Permanently fix the magnetic charge that orients these transistors, and you have a chip that functions as a ROM, or *r*ead-*o*nly *m*emory. Allow for the transistors to be reoriented, and you have a more interactive chip known as a RAM, or *r*andom-*a*ccess *m*emory.

Governed by a crystal clock that oscillates at a million cycles or more per second, electrons pulse through silicon circuits to produce the binary digits that are a computer's smallest and, in some sense, only unit of data. These pulses are measured in nanoseconds — one thousand millionths of a second — but to speed the process even further, computers clump together bits and shuffle them in packages of four, eight, sixteen, sixty-four, or, most recently, two hundred fifty-six bits at a time. The incremental sizing of these packages is due to the crystalline geometries of silicon. A group of eight adjacent binary digits clumped together and shuffled as one unit constitutes a *byte*. What makes a byte important is the fact that an alphabetic character can be represented by one of them. A kilobyte is equal to 2^{10}, or 1024 bytes, although in common parlance this number gets rounded off, in this case, to a thousand bytes, or 1K.

The KIM — an eight-bit microprocessor with 5K of RAM — shuffled eight electronic pulses at a time through a memory holding up to five thousand bytes. These numbers alone are not impressive. A game of Space Invaders operates on only a slightly smaller scale. But while the video game is frozen into perpetual intergalactic strife with ROM, or read-only memory, that in the KIM was wide open to random access. Within the limits of computer logic, it could be programmed to do anything imaginable.

After their summer at Professor Nauenberg's, the Projectors moved that fall into a house of their own. Doyne, Norman, and Letty had searched the county for someplace large enough to hold the first

Eudaemonic household. They finally found a rambling, wood-frame structure at 707 Riverside Street, a few hundred yards from the beach and just back from the levees that keep the San Lorenzo River from flooding the town built along its banks. The house and its barn had once presided over this stretch of riverbank as their sole occupants. But the acreage had long since been sold off for beach bungalows and condominiums, the barn was sagging, and the house itself was in need of cosmetic, if not structural, attention.

The Riverside neighborhood, in its democratic receptivity, held a smattering of every element found in this sun-drenched town of fifty thousand. Tourists unloaded children and baja chairs into cottages rented by the week. Retired couples turned their gardens into mini–citrus groves or Shangri-las overrun with bougainvillea and fuchsia. High-tech employees from Intel, after an hour's commute over the mountains, wheeled their Porsches into the front yards of otherwise unadorned condominiums. Other citizens, surviving somehow in an economy dependent on fish, Brussels sprouts, the university, a Wrigley's chewing gum factory, food stamps, silicon chips, and tourism, used their front lawns for planting snow peas, fitting skylights into Dodge vans, rigging Windsurfers, grilling vegetables over hibachis, or reading *Good Times*, the local newspaper whose masthead slogan is "Lighter than Air."

The flower-lined mall and cafés of Santa Cruz lay just across a bridge spanning the San Lorenzo, or one could stroll instead to the harbor, an expanse of blue water situated where Monterey Bay takes a final nip in the coastline before rejoining the Pacific at Lighthouse Point. Surfers off the Point shot the curl in Steamer Lane, one of the best surf breaks on the coast, while back in the quieter waters of the Bay one found a yacht harbor, a wharf with fishmongers selling the catch of the day, and a boardwalk complete with arcades and a roller coaster. The only incongruity in this pleasant neighborhood — which soon went unnoticed by its residents — was the screaming of riders on the roller coaster as they took the big plunge.

Besides its location, 707 Riverside had much to recommend it. A stone foundation, having already survived numerous earthquakes, supported a bank of stairs, a pillared porch, and a clerestory gable whose eaves and upturned roof made the house look vaguely like a Chinese pagoda. Despite its loftiness, the structure contained only one habitable floor, although one so extensive that it contained along its perimeter six bedrooms, as well as a living room, dining room, and kitchen built on a grand scale. The basement held two more rooms, with windows facing out onto a large back yard and the barn.

Then in her third year of law school at Stanford, Letty paid slightly over fifty thousand dollars for the house. "Norman and I were considering buying it ourselves," said Doyne, "but no one at the bank would give us the time of day. They thought Letty was pretty suspicious, too, until she produced her stock certificates. It was clear sailing from there."

Norman moved that fall from Portland to Santa Cruz and began his first year as a physics graduate student at the university. Letty came down from Palo Alto as often as possible. Juano, on being wiped out as a poker player in the card rooms of Montana, drifted back into town. The house filled up with other residents that included, over the years, scientists, teachers, lawyers, a pianist, a nurse, a volleyball coach, two Dutch film stars, and an Italian leftist from Milan. A way station for travelers and the headquarters of Eudaemonic Enterprises, 707 Riverside acquired the air of a commune, a physics laboratory, and a casino all rolled into one.

The Eudaemonic family fenced in the yard and planted a garden. They built tables and beds and bought other furniture at the Sky View Drive-In flea market. In a small white chamber off the front hall, Doyne set up the new computer in what came to be known as the Project Room. He lined the walls from floor to ceiling with shelves that he filled with shoe boxes containing electronic parts, technical manuals, spare chips, wiring diagrams, and other paraphernalia needed for assembling and programming the KIM.

The filigrees of silicon in a computer, its keypad, electronic circuitry, and clock are known as hardware. The second-level abstraction of a computer program — the set of instructions that actually endows the hardware with memory and logic — is called software. The KIM and other early microprocessor kits were shipped from the factory without software. And none existed.

As a computer the KIM was a tabula rasa, knowing nothing of language and numbers or their symbolic manipulation. This meant that it had to be addressed down at the brute level of electrons. In its state of near idiocy, the computer had to be spoken to in machine code, which is a combination of electronic bits no more articulate than one or two fingers wagged in the face of a gurgling baby. But from enough such repetitions even a dumb machine can wire its synapses into the pattern recognition of names and numbers. Playing with the computer ten hours a day, seven days a week, Doyne at the end of a month had taught the KIM how to multiply.

"First I had to teach it how to count to ten and back," he said. "The 6502 microprocessor manipulates data eight bits at a time, which means that it recognizes only 2^8, or 256 numbers between 0 and 255. This makes it complicated when you want to multiply 256 by 257. You also have to realize that for the early microcomputers 'multiply' was not one of the instructions. They understood only 'add' and 'subtract.' So you had to be able to multiply and divide any number that you were ever going to come up with by having the computer break it down into additions and subtractions of numbers between 0 and 255. And that's not a trivial thing to do. It took me a month.

"Then I had to teach the machine how to do sequences of operations with variable data, which is necessary for solving equations. More time went into teaching the computer how to handle logarithms, which are mathematical functions not expressed exactly by any finite combination of 'add,' 'subtract,' 'multiply,' and 'divide,' although they can be approximated by series of these steps. I needed to learn how to program a computer, and there was no software anyway; so it made sense for me to write my own routines. I was also concerned about speed. When I calculated the order of magnitude to produce something like a logarithm, I got worried, because it looked like a significant chunk of time. I knew I had to get the final answer in less than a second, and I was shooting for a tenth of a second. As it turned out, I brought it in under a tenth of a second with no problem."

Doyne drew up a "global plan" for organizing his work in the Project Room. He devoted part of every day to speculating on the theory of roulette, but the bulk of his time went to learning about computers. After soldering together the KIM and teaching it how to do arithmetic, he next had to instruct it in the logic of thinking.

"Microprocessors had been around for a couple of years, but the chip we were using hadn't been available for more than five or six months. The manual was freshly printed, with numerous errors, and no one had written any higher-level languages or assemblers or software support. I had to learn to program by directly entering binary numbers into the computer. A computer program is nothing more than a sequence of these numbers. The first of them is an instruction, which corresponds to one of the two hundred fifty-six things that the computer knows how to do. Now it understands what to expect from the next number, which will be either another instruction or a piece of data for it to act on. Further along in the sequence, the computer can make decisions, perform arithmetic calculations, or stroke its input-output devices. It can also handle 'interrupts,' which instruct it to jump around the program and look for other sets of instructions. The computer doesn't necessarily move through its program linearly. It can make loops and branches and jump among the numbers in some fairly complicated ways. And that's why computers can do nontrivial things."

As mapped out in his global plan, Doyne's work in programming the KIM represented only a small, if important, part of the project's attack on roulette. He also needed to solve the equations required for actually beating the game. "It was an ongoing process of figuring out how to make the predictions, along with careful derivation of the formulas and their testing through error analysis. It took me a fair amount of time just to come up with the basic idea of the program."

The game of roulette, with a ball revolving around a spinning disk, represents a model universe governed by the laws of Newtonian mechanics. Planetary ball circles solar disk until gravity sucks it out of orbit and pulls it down to stasis. The equations of motion governing this galactic drama are accessible to any freshman physics student who understands the meaning of $F = ma$. But various stumbling blocks have stood in the way of calculating this heavenly rendezvous and made it a classic problem in physics from the time of Pascal, who was Newton's contemporary, to the present.

Although the roulette cosmos works within the laws of gravity and planetary motion, its initial conditions alter every time the game is launched into play. This is comparable to the God in Newton's watchmaker universe reaching down fifty times an hour to tamper with the mechanism. To attain predictability in a world as fickle as this, one needs to clock the velocities and chart the relative positions of ball and rotor at the start of each game, and then calculate their eventual rendezvous within the ten to twenty seconds between the ball's cosmic launch and fall from orbit.

Humans not being fast enough, the only device competent for this act of celestial navigation is a computer. But when thought about in its totality, programming this computer becomes a daunting assignment. One has to derive and, if possible, solve the equations of motion governing roulette. The functions describing the game's individual parts have to be integrated into one comprehensive function, or algorithm, capable of making a split-second prediction. Other constraints make designing this computer even more of a challenge.

The machine requires, even in attenuated form, what are known as peripheral interface devices. This is the generic name for keyboards, terminals, joysticks, thumpers, microswitches, voice simulators, LED displays, and other means by which humans get information in and out of the central processing unit, or brain, of the computer. Through some such device, or combination of devices, the computer, at the start of each session, needs to be informed about the game's initial conditions. And then, on calculating the ball's trajectory and final point of contact, the computer needs a means of outputting its information. This computer with its peripheral attachments has to be battery operated, long lasting, concealable, silent, reliable, undetectable, and smart.

To predict the future and unfurl its mysteries out to the end of time, claimed the French mathematician and astronomer Pierre Simon, marquis de Laplace, all one needs to know are the position and velocity of matter in the universe at *one instant* in time. The same principle pertains to the mysteries of roulette. To become its Laplacian intelligence — its god of predictability — one needs to do four things. Predict how far the ball will travel before it falls off the track. Determine when the ball will drop from orbit. Predict how far the numbered cups on the rotor will have traveled by the time the ball falls into one of them. Add together the distance the ball travels and the distance the rotor travels to correlate their relative motion and the timing of their final conjunction.

Given the right equations and variables, four clicks tapped with the aid of a big toe will record everything required for roulette prediction by a Laplacian intelligence. Click the passage of a point on the rotor — the 00, for instance — in front of a fixed point on the rim of the wheel. Record the rotor's second turn in front of the reference point. Mark the passage of the ball in front of the same reference. Note the ball's second and subsequent revolutions past the marker. The more clicks for the ball, the greater the computer's accuracy in predicting its deceleration, but two alone, along with the two already entered for the rotor, will suffice.

"While deriving the roulette equations, I took everything I had already done on the campus computer," said Doyne, "which was written in the higher-level language of BASIC, and translated it into machine language. Then I had to think of ways to program the feedback stuff. What parameters did we need, and how were we going to set them? If you input data via microswitches under your toes, how does the computer unravel these clicks and decide what each of them means?

"It's no easy task writing a program to predict roulette, even when you know how you want it written. Getting the computer to perform the calculations took only a quarter of the time I spent on the program. The rest went into teaching it what the inputs meant and how to get information back to the outside world. Just making sure the computer didn't get confused about what it was supposed to be doing required a lot of care."

A good part of Doyne's program dealt with identifying the parameters that differ from one roulette wheel to the next. "There are basically five numbers you want to know. We devised a trial-and-error method for setting these five parameters. The computer is programmed in advance with ideal values. You click in data with your toes and compare them to what the computer expects to find. Then you fiddle with these predictions until you get each of them coinciding with reality."

For this fiddling process, Doyne divided the computer program into eight specialized domains, or modes, five of which were devoted to calibrating parameters for the ball, rotor, and tilt of the wheel. Two other modes were used as mathematical scratchpads for keeping histograms on how well the computer was performing. A final mode — holding all the adjustable parameters — was reserved for playing the game itself.

It is easy to get lost in a program like this, especially when you have to communicate with the computer through your toes. So Doyne devised something called a mode map. This diagrammed, in a series of interlocking loops, the relationship among the eight modes. A specific pattern of toe clicks steered the computer from one mode to another in a procedure that came to be known as "driving around the mode map."

Two big toes were actually required for driving around the map. The left toe looped the computer from mode to mode, while clicks made with the right toe incremented or decremented parameters. A complete tour of the map, with stops along the way for adjusting variables, took about fifteen minutes and ended with the computer being clicked into the playing mode. Once its parameters had been fine-tuned to the roulette wheel at hand, the computer's predictive power was uncanny. Even Laplace might have been surprised by the intelligence available to a big toe.

4

Radios from Other Planets

The computer can't tell you the emotional story. It can give you the exact mathematical design, but what's missing is the eyebrows.

Frank Zappa

If the Eudaemonic KIM was a mother computer destined for re-production, its first offspring was going to be, relatively speaking, a giant. Four by five inches — roughly a quarter the size of a printed page — the prototype roulette computer would itself be cloned into later versions no bigger than a library index card. There was a simple practical reason for starting big: transistors are easier to count when spread out over surfaces larger than the head of a pin.

Juano was nominally in charge of building the prototype com-puter, although Doyne and Norman helped design it, and Jack Biles assisted in putting it together. They decided as a general rule to christen their new computers with the middle name of the pri-mary builder. So the Project's first homemade computer, con-structed by John "Juano" Raymond Boyd, came to be known as Raymond.

Raymond caused a lot of trouble right from the start. Built with a microprocessor and chips bought off the shelf from parts houses in the Silicon Valley, it showed them they had a lot to learn about making computers from scratch. "We became a familiar sight to suppliers over in the Valley," said Norman, "as we accumulated a nice supply of burnt-out chips."

Even under the tutelage of Dan Browne, Juano had lost his shirt as a poker player. On returning to Santa Cruz, he was nominally looking for a job, but with black hair draped over his spectacles and halfway down his back, Juano — actually the most harmless of people — looked like a refugee from the Haight-Ashbury drug wars. After a job conducting telephone surveys, and another sorting parts for an electronics firm in Santa Cruz, he tried to capitalize on his B.A. in physics by selling himself in the Silicon Valley as a technician. While commuting to interviews on the far side of the Santa Cruz Mountains, he was able to shop in the Valley for chips, resistors, capacitors, diodes, crystals, and the other ingredients needed for making computers at home.

The microproccesor in Raymond was identical to that in the Keyboard Input Module. The major difference between the two computers lay in their memories. While the KIM stored its programs on cassette tape, Raymond needed something more compact. To come to life as a portable roulette computer, it had to incorporate a silicon chip for storing information known as a PROM, which is short for *p*rogrammable *r*ead-only *m*emory. To program this new memory chip, the Project required an electronic circuit known as a PROM burner, which can duplicate the memory on one chip — its orientation of 1's and 0's — and burn it onto another.

Doyne built a PROM burner onto the KIM. Juano drove over to the Valley for another handful of chips. They wired the components onto a circuit board and succeeded finally in burning the roulette program from the KIM into Raymond's memory. "At that point," declared Doyne, "Raymond was a computer in its own right and ready to go.

"In our spare time, Norman and I had been designing the hardware for the computer. Once you come up with a program, you have to imagine a circuit for getting it to work and a way to input and output data. We devised a plan for this and drew up schematics. But still in January and February, and even later into the spring, we were struggling to get Raymond running."

Computer troubleshooting is known as debugging, which Anthony Chandor, in *The Penguin Dictionary of Microprocessors*, defines as "the process of testing a program and removing faults. Ideally, a single phase in the development of a program in which the program is run with test data to test all branches and conditions that may exist in the program. Unhappily, debugging can often continue throughout the working life of a program."

Chandor only hints at the misery of it. He neglects to mention the fright of debugging a program for which there is no test data, or the appalling fact that bugs can exist in the hardware itself. Computers can be lousy with bugs, so thick with them that one suspects — in the ultimate nightmare — that bugs are shuttling back and forth at will from software to hardware. This is comparable in human pathology to the mind-body problem, where disorders in one realm delight in sneaking into the other. Some philosophers use this fact to argue against the Cartesian splitting of mind and body, and so too do the more existential hackers, when confronted with the mysteries of debugging, sometimes suspect the identity of software and hardware.

As Doyne bluntly put it, "Debugging Raymond was hell. I remember spending at least a month on it, and doing practically nothing but that. It was really very nerve-racking."

Norman's thoughts were no more fond. "We had a miserable time troubleshooting that sucker. We got really depressed at one point, because it looked as if the computer was just not working. So we built a separate machine that clamped onto Raymond and made it run through its program one step at a time. That's when we discovered we had left out a three-cent resistor. The problem was as simple as that. We popped it in and the computer worked fine."

With Raymond up and running, the Project jumped ahead to build its first casino-model microcomputer. "Raymond was our prototype," said Doyne. "It was meant as something to debug on. We thought it was too big to take into the casinos, although later it actually did get taken in."

It requires more than a handful of chips to make a computer. They are inert without electrons flowing through them. To generate and control this flow, the chips need to be wired into an electronic circuit made of transistors, resistors, and other components. Fragile and little bigger in size than the head of a pin, silicon chips come housed in plastic or ceramic cases known as DIPs, which is short for *d*ual *i*n-line *p*ackages. These resemble black centipedes with golden legs. Electrons flow up and down the legs, but only after the DIPs have been mounted onto circuit boards that hold them in place and organize the flow of current among them.

There are various ways to load chips onto a board. For Raymond, the Project had used a technique known as wire-wrap. The DIPs, after being mounted into sockets with pins of their own, had been plugged into a circuit board, and then a spider's web of interconnecting wires had been wrapped on the underside of the board from pin to pin.

"We wanted to make the second computer much smaller," said Doyne, "and we didn't think wire-wrap would work. It looked good for prototyping, but not for production. That's when we heard about a new technique, some kind of wonder wire whose insulation was supposed to melt off when you hit it with a solder gun. It did, but it also shorted out and melted all over the board. The wires were so ridiculously thin that they'd break if you sneezed on them. We ended up with a huge mess of Hansel and Gretel twine wound around the computer. We were jumping ahead too far when we built that one."

Raymond's first progeny — junked in the chip horde without having shown any signs of life — went unnamed. The Projectors returned to wire-wrap and started building another computer. By fitting the chips more snugly together, and clipping the pins on the bottom of the DIP sockets, they shaved a half inch or more off each dimension. After several weeks of soldering and debugging, they detected the first signs of life in the new computer, named Harry, after its paterfamilias, Norman Harry Packard. Raymond and Harry would take their places in this family history as the first Eudaemonic computers to cross the border and confront the roulette tables of Nevada.

Beyond building and programming computers, a third part of the Project's "global plan" remained, at best, fuzzy in outline. Thinking their presence in the casinos would be less suspicious with tasks divided between a data taker, standing near the wheel, and a bettor, positioned farther down the layout, the Project had opted for a two-person system using computers run in tandem. The data taker, while setting parameters and clocking the ball and rotor, would play penny-ante stakes, and might even choose to place losing bets. The second player, wired with a computer that received and decoded signals sent from the data taker's computer, would play the high-stakes game. Standing far from the wheel while racking up an obscene pile of money, the bettor could foil suspicion with any number of innocent disguises.

Chips of the gambling variety have a magnetism of their own. They draw energy like a short circuit. Stack enough of them in front of a player and crowds gather to stare. This is when the management gets worried and turns on the heat. At either elbow come the inquiring eyes of croupiers, pit bosses, shift foremen, shills, floormen, and dicks. They ply you with alcohol, distract you with questions, chew gum, sneeze, rub their crotches. "What," they ask themselves, "is causing this untoward good fortune? Is this guy on the level, or is he doing us dirt?"

While beneficial in terms of security, a two-person roulette system calls for some tricky engineering. It requires the communication of signals, by radio or other means, between one computer and another. Once transmitted, these signals have to be decoded into some humanly understandable form — such as an LED readout built into a pair of spectacles, or tones sounded in the ear canal, or shocks or thumps received on various parts of the body. The problems involved in designing a two-person system such as this — one demanding both computer-to-computer and computer-to-human communication — lie in the realm of electrical engineering. The only person among the Projectors with experience in this domain, coming from his childhood days as a TV repairman in Silver City, was Norman.

To begin with, Norman tackled the problem of intercomputer communication. He decided that a radio link, which was simple enough to build, was also too obvious. Large casinos are wired from end to end with detectors for bombs, cameras, guns, and other objects identifiable within the normal band of radio frequencies, and one could expect to find sensors in the Eye in the Sky above every roulette table.

Having vetoed a radio link, Norman explored other options, including ultrasonics, which are untraceable without special detectors, and an optical system designed around infrared lasers. Built into the heels of shoes, lasers could transmit signals through a sheet of infrared light invisible to the human eye. "Both ultrasonics and lasers worked perfectly well," Norman reported, "so long as no one was standing in the way."

It was Tom Ingerson who suggested the solution to Norman's problem. While on sabbatical in Chile, Ingerson maintained a steady correspondence with Santa Cruz in letters full of everything from philosophy to wiring diagrams. "I suggest you go the James

Bond route," he wrote, "and put a radio receiver in someone's tooth. It isn't hard these days, and one can send signals by biting down, and receive actual numbers to make the bets on by bone conduction, which cannot be heard by anyone else."

As an alternative to the James Bond route, Ingerson advised the Project to transmit the signals by means of Faraday, or magnetic, induction. This works on the same principle as a transformer, where current passed through a coil of wire alters the magnetic field and thereby creates a current in a second coil of wire. The flow of voltage through proximate wires is described by Faraday's law; hence the name for this kind of signal.

Unlike radio waves, which propagate through space in a radiation field, Faraday induction creates slowly varying magnetic fields that lose very little energy to the outside world. "They get weak so fast," Norman discovered, "that beyond ten or fifteen feet you can't detect them at all, even with the best of instruments. This was obviously a big plus. Our signals would be virtually undetectable to the Eye in the Sky, or anyone else in the casino not standing directly next to us."

Norman flipped through his textbooks on electricity and magnetism and sat down to build a set of transmitters and receivers. Little did he suspect that they would be the bane of his existence for years to come. He would spend thousands of hours troubleshooting the mysterious malfunctions that plagued their circuitry. What was intended as a bright idea for slipping through casino security would strain, in its complexity, even Norman's exemplary patience.

He was working simultaneously on the second of his two assignments: communication between computers and humans. One usually talks to computers through keyboards and visual displays, but for obvious reasons the Projectors required less obtrusive devices. For entering data into the computer, they had already replaced the keyboard with toe-operated microswitches. Now they needed some way to get information back out again. The options available to Norman included visual, audio, electrical, and tactile outputs.

He rejected the visual and audio as too obvious. Watches with LED read-outs or hearing aids sounding tones in the ear canal were both detectable. To examine tactile outputs, he got in touch with a company in Palo Alto that made devices for the blind. Their most

promising product was a machine that translated printed text into
buzzes readable by fingertips. But Norman found the machine too
fragile and inflexible about the kinds of voltage it required.

Having ruled out visual, audio, and tactile devices, Norman
turned to look at electrical outputs. "The perfect solution, we decided, were shockers, which would be totally flat and easily
concealed."

Electrical shocks sent to the body could be decoded as signals
identifying a particular octant on the wheel. Roulette wheels are
conveniently etched with black lines that slice their central disks
into eight pie-shaped wedges. The thirty-eight pockets on the
wheel are not evenly divided by eight; so each wedge contains between four and five numbers. A computer narrowing the outcome
in roulette to a particular octant, and allowing time to cover these
four or five numbers with bets, would give someone a killing advantage in the game — if Norman's shocks had not already done
mortal damage of their own.

Using the tops of tin cans coated with a "special medical conducting goop," he wired a ground electrode into the small of his
back and four shockers to various parts of his body: one on each
leg and two on the stomach. With combinations of shockers vi-
brating together, he could generate more than enough signals to
identify the eight octants and a ninth "no bet" signal.

"The idea was to send voltage through the system, causing a little sensation, presumably not an unpleasant sensation, although it
was hard to get the voltage right." This was Norman's mild-man-
nered way of admitting that the Project Room at one point resembled death row at Sing Sing, with human guinea pigs writhing on
the floor in the early stages of electrocution.

After giving up on tin cans, Norman ordered from Hewlett-Packard special medical sensors used for measuring EEGs and cardi-
ograms. "These," he declared, "didn't work worth a damn, even
after we shaved our skin and gave each electrode its own ground.
The current was too hard to control as it flowed over our bodies.
Once it entered, it could travel around and come out all sorts of
places. So after many months, we finally gave up on shockers."

In spite of the precautions that kept talk about it to a minimum,
the Project began to accumulate a group of interested friends.

Those knowledgeable about gambling offered advice on casinos. Others enlisted as potential players. And still more people, acquired by word of mouth and chance encounters, lent technical expertise. One of these was Jonathan Kanter, an electronics whiz who appeared like an answered prayer to solve the problem of the shockers.

Doyne was driving the Blue Bus to the university when he stopped one day to pick up a hitchhiker. In hopped Kanter. Slender, intense, his long brown hair roped into Jamaican dreadlocks, "he looked like a white Rastafarian," said Doyne, "with a rat's nest of hair that probably hadn't been combed in three years."

On the spectrum of computer freaks, Kanter was way to the left. He went barefoot and laughed like he was stoned all the time. A New Yorker who had dropped out of high school to go west, he lived in a garage filled with electronic gear and supported himself by building telephone blue boxes. Sleeping until two in the afternoon and then working until dawn, he kept the classic hours of a computer nut.

They started talking and Doyne asked Kanter, "What are you doing up at the university?"

"Roulette," Kanter said. Doyne was surprised to hear that he had got the idea from reading Thorp's book, and that he was now involved in building a roulette computer with a professor named Ralph Abraham.

"I met Ralph late in 1975," said Kanter. "At the time he was into building a machine for beating the Wheel of Fortune. With a simple analog device, he wanted to predict from the sound where the wheel would stop. The device gave you a voltage whenever it heard a click, and from that you could figure out the rate of deceleration. A drummer and bass man with perfect pitch had done the same thing merely by listening to the wheel.

"I had just turned twenty when I got involved enough with Ralph's project to go to Tahoe with a tape recorder. That's when I read Thorp and learned about using a computer to predict roulette. I ran to Ralph with the idea and he said, 'No, it's too difficult.' But we went to work on it anyway.

"He thought up the idea of stroboscopic sunglasses, which we tested with a strobe light and a bicycle wheel spinning at the same speed as a roulette wheel. We discovered the ball decelerates too quickly for this to work. Then I started building an analog machine that would make predictions from the sound of the ball as it travels around the track. The volume changes as it either approaches or moves away from you. I took my recordings to the Center for Research into Acoustics and Music at Stanford to run some spectrum analyses. I wanted to see if the *frequency* of the ball changed as it decelerated. One look at the spectrum and you knew this wouldn't work either. But the *amplitude* tests showed definite sound peaks. As the ball gets closer, its volume increases. So it seemed to me that what we wanted was a peak detector."

Unknown to Eudaemonic Enterprises, Kanter had rented a roulette wheel from the same game store on Market Street in San Francisco. He tried tracking the ball on videotape, but gave up when he discovered that his pictures were hopelessly blurred. He then turned to high-speed movies with a digital read-out flashing in the background.

"I was still thinking of working with sound, but I soon got hip to the idea that sound wasn't enough. Sound alone would allow you to follow only the ball. That's when I thought of radar and the Doppler effect."

Kanter ordered a Doppler radar device from West Germany: a Valvo MDX 0520, costing two hundred sixty-five dollars. Cheaper models used for home burglar alarms can be bought at any hardware store, but they have horn antennas in the shape of metal ears or funnels attached to their bodies. The Valvo was completely flat, with a printed circuit antenna made from windings of copper. The size and color of a hockey puck, the device produced via radar a direct measure of speed for an object moving either toward or away from it. It did this by means of the Doppler effect, which explains, in its most famous example, why a train whistle changes pitch as the train approaches and passes in front of you.

On hearing Kanter's story, Doyne told him about his own work in roulette and invited him to 707 Riverside, where the roulette wheel, along with the KIM, Raymond, and Harry, had been set up on a picnic table in his bedroom. Doyne was putting the finishing touches on the program while running the computers against the game in play.

Kanter was "really impressed" on finding this mini–casino and electronics shop in someone's bedroom. "The wheels I had rented in San Francisco were in terrible shape compared to theirs, which was new and shiny. They also had an array of different-sized balls, which I knew at the time to be important. Doyne showed me the measurements they had made with photo diodes mounted around the sides of the wheel, and this was clearly a better way to get data than my earlier attempts in San Francisco. They were the most together people I had run across."

On Kanter's bringing over his Valvo radar device, they found it worked perfectly on the wheel, where it could track the metal cups spinning on the rotor. It worked less well, though, in measuring the velocity of the balls, particularly those made of Teflon, which is nearly transparent to microwaves. Eudaemonic Enterprises decided to stick with the stopwatch system of switches and toe clicks. What this input lacked in accuracy, it made up for in versatility.

While most of its concerns were mathematical and electrical, Eudaemonic Enterprises also confronted problems that were psychological in nature. Given the distractions found in casinos, how could data takers train their toes for accuracy in clicking? And how were bettors going to keep cool in the face of casino heat?

To tackle the first of these problems, Doyne set up the eye-toe biofeedback machine and established a regular schedule of practice sessions for anyone hoping to play in the casinos. First prize for the winner of these Riviera Sweepstakes — that person displaying the best eye-toe coordination — would be a trip to the roulette tables of Monte Carlo.

Consisting of two parts, an infrared photocell directed onto the track of the roulette wheel and a toe-operated microswitch mounted into a pair of sandals, the biofeedback machine was wired into the KIM computer, which registered a toe click and a photocell "click" at each passage of the ball around the track. After comparing human time and photocell time, the KIM flashed the difference on an LED display. It also compiled running averages.

"There was quite a range of abilities," said Doyne. "The best people had errors within three hundredths of a second, while others couldn't get within a tenth or even a quarter of a second. This correlated strongly to how athletic you were, and men typically performed better than women. We did worse when we were tired and discovered on getting stoned that our averages went down with every toke. With alcohol, on the other hand, some of us improved after one drink. So we figured out the optimal amount of alcohol required. The key was to be concentrated but relaxed, like playing tennis."

Always a strong athlete, and later a professional volleyball coach, the winner of the Riviera Sweepstakes was Steve Lawton. With an average clicking error of less than three hundredths of a second, he was henceforth known to the Project as "Stevie the Toe."

Although still officially on leave, Doyne returned to the university that spring as a teaching assistant in the introductory course on electronics for physicists. "I had run out of money. My bank account read zero dollars."

"It was a lonely, difficult year for Doyne," said Letty. "He used up all his reserves during the time he was out of school. His financial reserves. His reserves of self-confidence and initiative. Anyone else embarking on the Project would have quit long ago, and those who knew what he was doing were bowled over by his persistence and drive."

Doyne had invested two thousand dollars in the Project and lent another fifteen hundred to Norman so he could finish college and start graduate school at Santa Cruz. "Norman," said Doyne, "had exceeded every limit everywhere for student loans. He was so in debt that for several years he wore braces on his teeth which were neither tightened nor removed because he couldn't afford a visit to the dentist. I finally dug into our electronics toolbox and used the same long-nosed pliers and clippers that had built Harry to remove the braces myself. It was really pretty simple. With a little more practice, I could have gone into business."

Teaching electronics was equally painless (Doyne gave the course lectures on microcomputers), and working at the university proved a good way to enlist bright students into the Project. Doyne kept an eye on five in particular: Marianne Walpert, Ingrid Hoermann, Mark Truitt, Rob Lentz, and Sandy Wells, the last of whom was hired that spring to rebuild Norman's radio receivers. The others would also eventually commit themselves to working for a slice of Eudaemonic Pie.

Being back on campus gave Doyne the chance to talk about roulette and computers with selected faculty members. Most of them had no idea why he had dropped out of school, and those in the know about the Project were close-mouthed to the point of paranoia.

"Everyone wanted to know what Doyne was doing," said Norman. "He gave them a line about how he was working on a secret money-making scheme that would release us from the fetters of the rat race. Most of the professors would nod their heads and say, 'Oh, yes, that means patent agreements. You have to treat these things with propriety.' But others were peeved about not being included in the inner circle of confidants. Whenever we did talk to someone about the Project, they were tickled pink by the idea of our pulling it off."

George Blumenthal, Doyne's former adviser, thought his work on roulette sufficient for a Ph.D. Other faculty members reviewed his equations or reflected generally on the problem of beating roulette. Throughout these discussions the name of Ralph Abraham kept popping up. A gambling system is ultimately only as good as its concealment, and Abraham was clearly the recognized expert in Santa Cruz on casino surveillance.

For a modest town of difficult access, Santa Cruz is endowed with a surprising array of luminaries who wander its forests and meadows like intellectual nabobs displaced from more traditional centers of culture. Loose in the redwoods at that time were Norman O. Brown, Herbert Marcuse, Gregory Bateson, and John Cage. The satyr among them, sporting a pepper-gray beard and bemused, if not sardonic, smile, was Ralph Abraham.

Professor of mathematics at Princeton, Columbia, Berkeley, and Santa Cruz, author of the pre-eminent text in classical mechanics and five other books, specialist in nonlinear analysis, dynamical systems, morphogenesis, and pattern formation, Ralph Abraham arrived in Santa Cruz in 1967 at the age of thirty-one and dropped acid for the first time. "That was the turning point for me," he said. "I began my life on the road, the search for the miraculous, and also my life of crime."

On his first day in Santa Cruz the yet-to-be transformed Abraham was instructed by his friend Page Stegner to look up Fred Stranahan. "I drove over to Jim Houston's house in a Shelly Cobra that I had rented from Hertz," said Abraham, "and Houston told me I'd find Stranahan out at the Barn in Scotts Valley."

The home of the Merry Pranksters, the Barn was painted in psychedelic colors, lit in black light, and filled with space music made by the Sons of Eternity, who played instruments shaped like pornographic sculptures. "There were three hundred people in there," said Abraham, "all tripped out on acid — kids, dogs, everybody. I had my first trip there, and then I saw what was happening."

On moving to Santa Cruz, Abraham bought a twenty-four-room Victorian mansion on California Street and began "listening to radios from other planets. I gave my life over to the *I Ching*. I began traveling, spending a year in Europe, sleeping on floors, in opium dens and train stations, and then I thought I was seasoned enough to go to India. I studied the Vedas for seven months and supported myself by giving math lectures. I looked into all the available forms of mysticism: Gurdjieff, Sufis, astrology, and politics, too. But I decided, rather than mysticism or politics, that 'the way' for me was to travel, that it was better than belonging to a group or meditating or having a guru.

"There was a lot of pressure from Governor Reagan and the chancellor of the university to have me quit; so they were glad to see me go traveling. On getting back from India, I made a survey of all the professions that would support my way of life. There was a list of criteria by which I evaluated them: flexibility, good working conditions, profitability, and transportability around the world. I chose gambling as my profession."

Abraham sold his house and moved into the St. George Hotel, a way station for transients down on Pacific Street. He set up shop in the Catalyst, a bar then located in the lobby of the St. George. "I ate all my meals there while practicing card counting for weeks on end. I already knew Thorp's book. I read it again, learned the system, and practiced with flash cards in the Catalyst, until I later worked out a scheme with a projector and slides showing two hundred different possibilities. Then I headed for Nevada."

Abraham and I are drinking chai and eating tumis — stir-fried vegetables — in a Santa Cruz café called India Joze while he recounts this part of his life. Bespectacled, tanned, with a prepossessing forehead and piercing dark eyes, he looks like a professorial guru conscious of the veil of maya, but tenured into it. In preference to tweeds or saffron robes, he wears a multipocketed vest and cowboy snap shirt. There is a long pause in the narrative before he resumes.

"I lost continuously. Five thousand dollars altogether. Then I discovered Larry Revere's book. No doubt about it, it describes the world's finest blackjack system. I began losing faster. Then I took my hundredth look at Revere's book and found on the last page a sentence that says you can't learn a system out of a book.

"I headed for Las Vegas and found Revere. I gave him my last hundred-dollar bill and said, 'I want to be your student.' He was a millionaire several times over. It was a power token, an act of humiliation. I became his star student. He taught me how to walk, how to dress, how to sit down and leave a table, how to shuffle money from one pocket to another. He was an acting teacher, a master of disguise.

"'Revere' wasn't his real name, and the name before that wasn't real either. After his first lesson, I began winning, but I knew he was holding out on me. I wanted his knowledge. He had bought a casino and was playing both sides of the business, and by then he needed a secretary; so I inserted a friend of mine from Santa Cruz into his life. I phoned her up and she flew to Vegas to go to work for him.

"But as I said, from the day I met Revere I became a successful blackjack professional. I installed myself in Tahoe and began making a couple of thousand dollars a month. Playing two hours in the morning and two in the evening, I cleared twenty dollars an hour. I could have made five times as much in Vegas, but I couldn't stand the place.

"With a disguise I could play the same casino for weeks on end without their knowing I was winning. The acting is all in your hands." Abraham opened his vest and showed me the extra pockets sewn inside. "You have to shuffle chips in and out of these pockets. Knowing exactly where they are, you want to look as fuddled as everybody else. You're sitting at a table with people who are losing, and if they weren't losing you wouldn't be winning. While putting a lot of chips *into* your pockets, you have to make it look as if you're pulling them *out*.

"You also need a strategy for cashing your chips. You can't do it in the same casino; so you take advantage of the courtesy most of them extend in cashing each other's chips. You memorize shift changes, so that losses get reported on different shifts, and then you drive around town in a certain pattern to pick up your winnings. These techniques are known as 'money management.'

"While you're playing you have to count all the cards in twenty-six seconds. Twenty-eight seconds and you lose. The game just goes right by you. There are a hundred hands per hour, thirteen cards per hand, over a thousand cards per hour, and you can't afford to make *one* mistake. I used to practice for an hour in the morning before hitting the casino. You have to count cards out of the corner of your eye while talking to the guy next to you. The place is noisy, like an airport on Sunday night. The dealers, pit bosses, cocktail waitresses know a hundred ways to make you lose your concentration. And once in a while, when a dealer suspects something, he'll try to trap you. He'll shoot a card traveling at forty miles per hour straight for your nose, and you can't catch it, or he'll know what you're doing.

"Back then there were maybe two hundred professional blackjack players. Now there are thousands. Ken Uston took over from Revere when he died and introduced the concept of team play, with groups of a dozen going out on the road for a year. Stanford Wong worked out a strategy based on walking from table to table, called 'wonging.' He learned how to watch the dealers' faces for telltale signs, or 'tells,' which tipped him off on when to bet.

"With the proliferation of schemes there was a war going on between the casinos and the professional gamblers. Revere himself got co-opted when he bought a casino. Soon it was impossible to tell who was whom, especially when the casinos started pushing schemes themselves. They understand that nobody loses money faster than a card counter. One mistake and you're lost.

"Every technician over in the Silicon Valley thinks of himself as a gambler. On weekends he drives to Tahoe to count cards, or putters around in his garage wiring semiconductors into a gambling system. But in this business either you're a professional or you're nothing. The casino world is an airtight community, an alternate reality. You live in the hotels and play downstairs in the air-conditioned casinos and never go outside, because nothing outside is as interesting as inside.

"You see the casinos wheeling carts around to pick up the cashboxes. Hour after hour they're raking it in. You're sitting next to losers who don't leave the table until they're broke. You see their wives pulling on their sleeves, saying, 'Honey, don't do it. That's our bus fare.' And then it's gone and you have no idea how they're going to get back to wherever it was they came from.

"The gambling is all run by the Mafia, and it's their best business. Who knows what makes people do it? It's an animal instinct, an atavistic trait, a disease. Sitting there in the middle of it is like being on Forty-second Street at Grand Central Station. You see every kind of person in the world. They drop all pretense. They mouth their incantations over the roll of the dice or the card on top. It exceeds all other levels of reality. It's naked reality. I mean the casinos are so bad and greedy — screaming about every nickel they lose, while cheating and robbing people blind — that to beat them at their own game is a white knight operation."

After Reagan moved up to national politics, Abraham returned to California. "I'm a contactee, an astral projector into alternate realities, a listener to radios from other planets, and it was a bit odd," he said, "being back at the university. My specific goal is to revolutionize the future of the species. Mathematics is just another way of predicting the future."

The three of them later became good friends, but Doyne and Norman hesitated before talking to Abraham about the Project. Who knew what alternate reality he might make of it? Indeed, their first meeting went rather badly. "I studied their algorithm, and it looked like it would do," said Abraham. "I reviewed their statistics, the fatigue factor, the equipment they'd put together, and it too looked like it would work.

"But they had no concept of casino security. They thought they could swap battery packs and equipment in the hotel toilets, which are monitored, of course. I thought the consequences could be severe if they were caught. Ken Uston was in the hospital at the time, having reconstructive surgery done on his face.

"I tried to warn them in that first meeting about the dangers of detection. They said their technology was too sophisticated. I believed they'd give themselves away. They thought I was paranoid. I thought they were conceited. I told them their best bet was to sell the computer to somebody who knew what he was doing."

5

Debugging

Things are not as simple as they at first seem.

Edward Thorp

After a year of building computers, transmitters, receivers, shockers, and the biofeedback device, Eudaemonic Enterprises paused for a spring party also meant to celebrate Letty's graduation from Stanford Law School. Entitled "Come as You Are in 1997," the party was projected as the twentieth class reunion for the Class of 1977 — a time warp into which costumed participants would slip for a peek at themselves on the brink of the third millennium.

Norman made a 1997 calendar to hang in the hall and unfurled a "Welcome Class of 1977" banner over the front door. (The banner drew several confused participants from off the street.) All the rooms in the house were made over and dedicated to one form or another of pleasure. The living room was wired with strobe lights and turned into a discotheque. A movie room showed nonstop Abbott and Costello films and abstract color field productions by Larry Cuba. An R&R room was lit with candles, and another chamber had been converted into a tactile room modeled on the Exploratorium in San Francisco. Completely darkened and lined with mattresses, the room and its walls were covered with everything from salami to fur.

Doyne's bedroom, renamed the Neural Stimulation Room, was converted into a shrine commemorating the excesses of the 1960s. Set up on an altar surrounded by signs referring to the days "when hippies used to sear synapses and pulverize pain with mind-expanding drugs" was a Jacob's ladder, with electric current arcing up it, and a bowl full of Kool-Aid punch.

The Neural Stimulation Room also featured Doyne's biofeedback device, which was appearing in public for the first time as a reflex tester. Set up on a bicycle wheel and operated by hand rather than toe switches, it otherwise employed the KIM computer and the same system of photocells wired into an electronic circuit.

Other equipment from the Project surfaced during the party. Dressed as a 1997 hipster, a bearded Norman wore a red headband, a burnoose that floated around him like a dress, and an LED necklace made from flashing diodes and a prism. "That's an example of an idea from which I could have made millions," he said. "But, alas, it's just one of those things I never got around to marketing." Appearing as Tom Terrific, Doyne sported a red leotard and cape, with a Peruvian medallion around his neck and a metal funnel for a hat. A light bulb on top of the hat, activated by one of the Project's toe-operated microswitches, lit up whenever he had an idea.

Many people pictured themselves in 1997 with extra organs and mutations. One guest showed up with a fully developed third eye made from a roulette ball. The first photosynthetic human came covered with green veins and leaves. Wearing a loincloth and burlap sack, with fetishes decorating his neck, Dan Browne arrived as a post–World War III cave man. Letty zipped around on jet-powered roller skates. Bruce Rosenblum, a physicist at the university, wore a Mexican cut-away jacket and a cone on his head covered with Maxwell's equations. Another professor, aged with charcoal creases and false breasts, came as Tiresias. Juano, dressed in a white robe and carrying a silver wand topped by a flashcube, kept himself enveloped in a cloud of gaseous nitrogen.

Many danced until dawn, while others emerged sheepishly from the tactile room late the following day. In the course of the evening, the reflex tester flunked everyone. "There was no doubt," said Norman, "that a lot of stuff consumed at the party was bad for your motor coordination."

"After Doyne finished programming the computer," Norman confided, "we thought it would be a matter of weeks before the money started rolling in. Once the program was finished, let's face it, that's the main thing. The Project was done in principle."

For an impeccable optimist like Norman, Project Rosetta Stone was already a fait accompli with roulette wheels knocked off all the way from Monte Carlo to Macao. "It took us a while after that before we actually got in the casino, but as far as our state of mind went, all we had to do were a few little things and *tsweeet*" — he makes a noise with his tongue expelling air through his lips — "we'd be in there, no prob."

Doyne was equally buoyed, until he became suspicious late in the spring that something somewhere was not working right. Word having gone out that a trip to Nevada was imminent, Jack Biles had come down from Oregon for a marathon session of hardware building. But beyond the usual problems in getting hardware soldered together and running, Doyne began to suspect that something might be wrong with the computer program itself.

"The big push during the spring and summer," he said, "was to get the program to predict roulette in real time." He went back to plotting histograms, filling sheets of paper with graphs showing the frequency with which the computer's predictions matched the ball's actual behavior. "I remember sitting there and taking histogram after histogram and not having any advantage. I got worried enough at that point to develop three alternate systems for prediction."

Doyne thought the problem with the computer program might lie in the fact that roulette wheels possess varying degrees of tilt, some being relatively flat, most being tilted several degrees, and others resembling the *Andrea Doria* ten minutes after the final call to abandon ship. He wrote algorithms to cover these varying conditions and programmed the KIM to play roulette with three different sets of equations.

Unbeknownst to Doyne, two of these three algorithms had already been identified by Edward Thorp when he and a partner — mysteriously left unnamed for many years — tried but failed to implement a computerized roulette system in the early 1960s. One reason for his difficulties lay in the fact that Thorp's system possessed a limited number of adjustable parameters. The player, for instance, had to guess the exact number of revolutions remaining before the ball fell off the track. Another reason for Thorp's disappointment — one entirely beyond his control — stems from the fact that the microprocessor had not yet been invented. Carrying precursor technology into the casinos, Thorp had had to work with approximations rather than precise equations, which, even if they had been known to him, would have been unsolvable by his computer.

Although Thorp's roulette system had been mentioned earlier in *Beat the Dealer*, its details were first revealed in a technical paper published in 1969 in the *Review of the International Statistical Institute*. This is not a journal commonly read by either physicists or gamblers, and few among the latter would have understood Thorp's equations anyway. So even after this breach of secrecy, the theory of roulette remained esoteric knowledge.

"Our basic idea," wrote Thorp — referring to himself and his unnamed partner — "was to determine the initial position and velocity for the ball and rotor. We then hoped to predict the final position of the ball in much the same way that a planet's later position around the sun is predicted from initial conditions, hence the nickname 'the Newtonian method.'"

As too many variables in the game lay beyond the scope of Thorp's linear approximations to its nonlinear equations, he and his partner abandoned the Newtonian method and developed an alternate approach that they called the quantum method. This took advantage of imbalanced roulette wheels and the fact that even a slight amount of tilt greatly simplifies one of the variables in roulette prediction: locating the point on the track from which the ball will fall and begin its spiral down to the rotor. On tilted wheels the ball races around the track at varying velocities. It alternately slows down and speeds up as it approaches and passes the high side of the wheel. Under these conditions the ball tends to come off the track when slowed for its climb up to the high side. Once over the hump and gaining in velocity, it hugs the track through a stretch Thorp called the forbidden zone.

The presence of tilt in a roulette wheel also adds a neat twist to the physics of the game. It allows for quantizing, or clumping into discrete sets of values, the position and velocity of balls coming off the track. Thorp explained the logic of the quantum method as follows: "Suppose the ball is going to exit beyond the low point of the tilted wheel. Then it must have been moving faster than a ball exiting at the low point, so it reaches its destination sooner. But it has also gone farther, and the two effects tend to cancel."

With these differences offsetting each other, Thorp realized that balls exiting from a discrete section — or quantum — of the track will all tend to strike the rotor at the same point. The more tilted the wheel, the "more sharply bunched or focused" will the balls be when they hit. Estimating that more than a third of the roulette wheels in Nevada had the required tilt of at least two degrees, Thorp calculated an advantage for the quantum method of over 40 percent. This is a tidy return on an investment that can be made and paid off every minute and a half!

Because of its relative obscurity, Doyne read Thorp's essay — on learning about its existence from Ralph Abraham — only after developing his own roulette algorithms. Having independently arrived at the same conclusions, including the importance of tilt for predicting the outcome of the game, Doyne nonetheless adopted Thorp's terminology. He appreciated how nicely it recapitulated the history of physics. "In the Newtonian method you assume there is a continuum of positions from which the ball can come off the track. Newton thought that all of physics could be described by such continua. When quantum mechanics overthrew the Newtonian picture, you could no longer assume the existence of a continuum, and what you had instead was matter broken up into quanta, or indivisible chunks."

By the time he read Thorp's article, Doyne had already organized the Newtonian and quantum methods into equations, which he solved for the first time with the aid of the Project's digital computer. He had also developed a third differential equation for describing roulette. Called the post-Newtonian method, it was meant for use on wheels intermediate in tilt between the flat and the rakish.

"At one point I contemplated writing an article on roulette algorithms and the physics of roulette balls for *Physics Today*. I imagine," he said with a smile, "that I'm the world's expert on the subject."

Working with the KIM computer in his bedroom, where it was set up on the picnic table next to the roulette wheel, Doyne began to get worried when he could get *none* of his three equations to work well in real time. They looked good on paper, but up against the wheel they gave only a slender advantage.

"I was being overly ambitious," he said. "I wanted to beat every possible roulette wheel. Some are going to be very tilted, some flat, and a lot in between, and the idea was to walk up to any one of them and be able to play.

"But I began to realize how long this thing was dragging on. I couldn't afford to spend months developing fancy algorithms to predict roulette when I didn't even have one that would work on highly tilted wheels, which are the easiest of all. Late that spring I got really nervous that the whole thing was just not working, that there was a flaw in the basic idea that we weren't taking account of. Maybe the ball was bouncing around on the track too much. It was jittery and somehow blowing the predictability.

"We still had data stored in the campus computer; so I went up to the university to try out the algorithms. No matter how I fiddled with the equations, the data on campus just didn't seem to work right. I did several experiments and got really depressed. It looked as if it might have been a fluke that we had gotten any advantage at all. At that point, I knuckled down and said, 'O.K., let's just go for the simplest thing. Tilt the wheel like hell, enter two clicks, and see if we can predict where the ball is going to fall off.'"

Doyne spent the entire summer reprogramming algorithms that failed to beat the roulette data stored in the campus computer. The Project at this point came to resemble an intensive language course in cuneiform. Written in floating-point binary arithmetic — the machine language understood by the KIM computer — Doyne's roulette program had grown to four thousand instructions. Reproduced longhand, it took fifty pages of binary numbers to label every location in the program. Each label, or address, in this string of numbers represented eight bits. The on-off orientation of a bit can also be represented by an electronic tone sounding either "high" or "low," which allowed Doyne to store these fifty pages of numbers in a tape recorder. When played back, these tones, sounding like the gibberish of a speed freak gone astral, filled ten very long minutes of tape.

Without high-level languages or other aids for traversing the fifty pages of a machine code program, Doyne struggled through them on guts alone. In the beginning he had had no choice. Fresh from the factory, the KIM got programmed in machine language, or not at all. Software tools known as compilers and assemblers can gather together machine code instructions and greatly simplify the process of programming a computer, but by the time hackers had developed them for the 6502 microprocessor, Doyne, strapped for money, decided against the investment.

On passing the bar exam, Letty took a job in Los Angeles at the Center for Law in the Public Interest, where she was offered the chance to work on the environmental and political cases that interested her. "We drove to Los Angeles and stayed with friends for a few days," Doyne said. "We wanted to check it out and decide if human beings could survive there. And then later in the summer I drove Letty down with all her belongings and helped her find a place to live."

On returning to Santa Cruz, Doyne sat down to look again at the roulette program. It was still not working right. Something, somewhere, was preventing it from attaining the accuracy it theoretically possessed.

At Ralph Abraham's suggestion, Doyne phoned Edward Thorp, who was then teaching at UC Irvine. They discussed casino security, and not the technicalities of the program itself, but this was the first of several brief encounters that Eudaemonic Enterprises would have with Thorp.

"Ralph convinced us," said Norman, "that Thorp was basically on our side in wanting to see the casinos beaten. He wouldn't give us away; he wasn't a casino man. Ralph also thought that if there were other systems being developed, Thorp would be the most likely person to know about them. We were interested in finding out if we had any competition. We also wanted to hear why Thorp had quit, if he *had* quit, and what the story was."

Thorp reassured them that a system such as theirs was workable in the casinos and that he himself had not met with any undue suspicion while using his computer in Las Vegas. He briefly discussed the reasons for his limited success, which he ascribed to hardware problems. "But he was vague," said Doyne, "about whether he or other groups were working on roulette — quite vague."

Back from his year in Chile and a trip around the world, Tom Ingerson showed up in Santa Cruz late that summer. While jogging several miles a day on the levees along the San Lorenzo River, he and Doyne talked about bugs in the program and other problems in getting the radio receivers to work properly. Ingerson came up with several helpful ideas, including a scheme for making the computer program "smart" enough to filter errors out of its signals. But his relationship to the Project was basically ambivalent. His sister and brother-in-law lived in Las Vegas, and he had spent a fair amount of time with them out in the desert. He had watched Len Zane dabble in card counting and then lose his nerve one day when a pit boss at the Sahara put his hand on Zane's shoulder and said he should take his business elsewhere. Like Ralph Abraham, Ingerson thought the consequences could be serious for someone discovered in the casinos wearing a computer.

In the meantime, Doyne was spending long days laboring alternately on the KIM and the university's PDP 11/45. "By the end of the summer," he said, "I still couldn't get the program to work right."

He went back to look again at the feasibility studies done the previous summer. These measurements of the game in play had been fed into the university computer and then modeled into the simulations on which he had based his algorithms.

"The data looked positive at the time. But I had queasy feelings about them, and my feelings had gotten queasier that spring when I went to campus and modified the programs for the post-Newtonian method. I realized that our data base wasn't big enough. I had the feeling we were fudging too many things. None of the methods seemed to work better than any other, which made me suspicious.

"So I wrote a special program that translated the hexadecimal floating-point binary arithmetic of the KIM — that's base sixteen with binary, or base two, exponents — into numbers that were readable by the campus computer. I transcribed the times from one computer onto the other and had them displayed in series, which I then compared to the times recorded by Norman's clock."

Doyne discovered a number of surprising things, including the fact that he had occasionally messed up his calculations by spinning the roulette ball in the wrong direction. But this alone didn't explain the problem, and he set to work ferreting other bugs out of the KIM's program.

"I thought we had made successful histograms as early as May or April, although I later realized these were statistical fluctuations. They were poltergeists. False starts leading nowhere. Several times I thought I had all the bugs shaken out, but the program still didn't work. So I'd go up to campus, run the program through, check it point by point, locate a couple of mistakes, and go home to make more histograms with the KIM and roulette wheel. I finally found a mistake in the last stage of the program, where it actually computes the answer. After I cleaned out the garbage, I was in utter despair when the program *still* didn't work."

At that point Doyne had no choice but to go back to the beginning of the Project and start all over again. He attached the photocells to Norman's electronic clock, set them up on the wheel, and began duplicating the data gathered the previous summer. Except that this time, alongside the clock, he ran the KIM, which allowed him to double-check whether clock and computer were synchronized.

Doyne soon discovered the cause for his summer's grief in debugging the program. "Norman's clock was out to lunch. It was messing up the original data all the way through. Probably due to hardware problems, the clock was making errors in recording the times. But these were *exactly* the kind of errors you could easily miss, because they were never very large. They averaged five hundredths of a second, which is just enough to simulate human error all the way along."

After dumping out the wrong and putting in the right times, the predictions on the KIM and the university computer "agreed every step of the way. I rushed home to tell Norman about it and we sat down at the wheel. I motored around the mode map, set the parameters, and clicked into the play mode."

To make way for a piano, the Project had been shuffled by then into a small chamber behind the kitchen. The room was crammed with the roulette wheel on its picnic table, shelves full of components, an oscilloscope, Raymond, the KIM, and the biofeedback machine. "Norman and I were really packed in there along with the roulette wheel and all this other stuff. The KIM and Raymond were balanced around the wheel, with wires running everywhere.

"To get all the data recorded at once and the wheel spinning at a reasonable speed takes some work. Norman and I sat down and got cranking. We were working as fast as we could, spinning the wheel and writing down data and graphing the outcome. We weren't even stopping to look at the histogram.

"We must have logged eighty runs. We were locked into the middle of this thing, until, finally, I turned to Norman and said, 'Let's take a look.' We held up the histogram, and there was a band of data points running straight up the center of it. The computer was doing just what it was supposed to do. We got really excited. We jumped up and down and hugged each other. This was our big breakthrough. A year and a summer after starting the Project, we finally had concrete proof that we could beat roulette."

A strawberry blonde with an impish smile and a flair for androgyny, Marianne Walpert presided over the Eudaemonic Halloween party that fall as Dionysius incarnate. Sheathed in a white robe and garlanded with a crown of laurel leaves, she walked among the guests serving laughing gas out of a plastic garbage bag. She had brewed her concoction in the Project Room, after having turned it for the night into a chemistry laboratory complete with Bunsen burners, Erlenmeyer flasks, pipettes, and other equipment needed for slow boiling and filtering the night's refreshment.

As a physics major at the university, Marianne breezed through her courses on curiosity alone. An argonaut into uncharted realms, psychic and physical, she also knew how to make friends and put people at ease. Ralph Abraham began frequenting the Riverside house after Marianne moved in, and another friend of hers, Alix Youmans, figures for an incandescent moment in the history of Eudaemonia.

Doyne had been Marianne's teaching assistant in Physics 6A, the basic course for first-year students. "He was very mysterious around school," she said, "and it was only a couple of years later, when he invited me over to dinner to see if I wanted to move into the house, that he took me into his room and explained what was going on. It had been such a deep secret that I didn't know what to expect.

"When I saw the roulette wheel and a roomful of electronic equipment, with chips and wires covering everything, I was totally amazed. I couldn't believe it. We talked about the Project for hours. I wanted to know how it worked and what they had figured out, because I was skeptical. I mean no one wins at roulette. But it seemed like a great idea to rip off casinos who get so much pleasure out of ripping off everyone else. I thought it was wonderful, and I couldn't wait to get involved."

After moving into the Riverside house, Marianne took off on an African trek. She flew to Paris, hitchhiked to Marseille, caught a boat for North Africa, and traveled overland across the Sahara from Tunisia to Cameroon. Having met Dan Browne only once before in Santa Cruz, she had convinced him to join her for the last leg of the trip. Flush with poker winnings, he was game for the adventure. Driving in a truck caravan, with one truck full of spare parts and others full of gasoline and water, it took them a month to cross the Sahara.

At the end of the summer, Marianne returned to Santa Cruz with a new friend whom she had met on the airplane, Alix Youmans, a thirty-year-old Parisienne in the midst of getting divorced from a wealthy husband in San Diego. Elegant and smart, she was versed in hiding her brains behind a patina of talk about astrology, est, and other psychic fads. Doyne imagined that Alix, with her French accent and wardrobe, could play perfectly the role of rich socialite accustomed to winning and losing large sums at roulette.

Alix began by lending herself to the Project's electroshock experiments in which current, run through electrodes and conductive cream, flowed indiscriminately over her body at levels often painfully high. This method of outputting data from the computer was later abandoned when Jonathan Kanter suggested they switch to a tamer form of mechanical output from solenoids. A solenoid is a little thumper that can be fine-tuned to vibrate against the skin at different frequencies. Activated by a magnetic field, it consists of a metal plunger that bobs up and down in a cylindrical coil of copper wire. Three solenoids, placed on adjacent parts of the body and vibrated at varying speeds, could transmit the nine signals needed to predict the outcome in roulette.

For the new solenoid system, the Projectors built three little thumpers into a metal plate, which was fastened under a belt and worn tight on the stomach. Translating local perturbations over the duodenum into a betting strategy took some practice, but otherwise the only problem with the solenoids lay in finding a way to hold their metal plungers in place. They bounced around like Mexican jumping beans, and would have flown around the room without something either tying them down or covering them.

"We experimented with Saran Wrap and Band-Aids," said Doyne, "but the Saran Wrap punctured too easily and the Band-

Aids gummed up. We needed something sensitive yet tough enough to withstand these repeated vibrations. We finally discovered the perfect solution: condoms held in place with radiator clamps. We made the early trips to Nevada with a large supply of rubbers and clamps."

In preparation for a trip to the casinos by December, Norman, Marianne, Doyne, and Alix got swept up in a flurry of hardware building. They manufactured the transmitters, receivers, and antennas needed for a radio link between data taker and bettor. They developed wardrobes and disguises and a coded language for emergencies, in which computers were called *brains*, batteries *energy*, and wires *nerves*. A remark about *alpha waves* meant that the *brain* was up and running, while *My nerves are shot!* referred to broken wires or a short circuit in the system.

On the afternoon of December 7, 1977, Doyne and Alix loaded Raymond and Harry, along with the transmitters, receivers, and toe-operated microswitches, into the Blue Bus. Doyne drove into the Sierra Nevada on Interstate 80, his destination being the casinos just across the Nevada line. Caught in a heavy snowstorm without tire chains, he and Alix barely made it over Donner Pass and down to South Lake Tahoe late that night.

In a laboratory notebook — the kind with black and white marbled covers and cloth tape binding — they recorded the trip's significant events in two columns. Marked "Left Page" and "Right Page," these columns were supposed to correspond in function to the two hemispheres of the cerebrum. The left page was subtitled "Journal (Life of a Gambler)," while the right page was reserved for "Technical Events."

The Journal for the first day records how Doyne and Alix slept in the Bus and woke late the following morning. "Not too cold, but damp," it said, before noting to "fix heater in the Bus, get a complete set of window curtains, fix window, fix leaks in roof."

The Technical Events are equally dour. They begin, "Electrical outlets in gas stations are very handy for soldering loose wires," and go on to record a chronology in which Doyne spent the entire day driving from one gas station to another fixing loose connections.

Alix, in the meantime, began another journal, this one in a black three-ring notebook. It recorded for each casino in Tahoe — and later for most of the casinos in Reno and Las Vegas — the layout of their roulette wheels, the tilt of the wheels, the names of croupiers, and various notes on shift changes and other pertinent data. The notebook also held financial records for each casino. These recorded the size of the bank used for betting, dates of play, numbers of trials, and money either won or lost.

With their radio link not yet debugged, Doyne decided to give up on the two-person system and play solo as both data taker and bettor. Under a loose-fitting sweater, he wired himself from head to foot with the computer under one armpit, battery packs under the other, the solenoid plate on his stomach, and toe switches down in his shoes. Needing ample room for the switches, he had built them into the kind of high-heeled footwear favored by pimps on Eighth Avenue. The shoes were treacherous for walking in snow, of which there was plenty that winter in Tahoe.

Doyne began by buying into a game at the Cal-Neva Club. He had finished setting parameters on the computer and was ready to play when a number of electric shocks, caused by short circuits in the wiring, forced him into the toilet. "I had a lot of problems with getting shocked. I would start sweating, and the sweat would cause short circuits, which would make me sweat even more. So on several occasions I wanted to rip off my sweater and throw it away."

With too much snow in the parking lots for Doyne to navigate in his "pimp shoes," and too little action in the casinos, he and Alix drove that evening to Reno. They parked on the edge of town and fell asleep in the Bus. Late the next day, Doyne wired himself into the computer. Walking into Harrah's, the poshest casino in town, he played another solo session while Alix collected data on roulette wheels — until the Project met with a near disaster.

Doyne had bought into the game and finished driving around the mode map when short circuits forced him away from the table. It was too cold to work outside at gas station outlets; so he carried with him in a book bag a few rudimentary tools for fixing the computer.

"I locked myself in a nice, cosy toilet down in the basement of Harrah's. The computer was open on my lap, along with an ohmmeter I was using to check voltages along the lines. It's hard to fix a computer with nothing but an ohmmeter. It's like trying to rebuild the engine in your car with a pair of pliers and a screwdriver. I had a pocket full of extra chips and was just about to insert one into the computer when a security guard poked his head over the toilet stall. He was a young guy about my age with a mustache.

"'Hey, what are you doing down there?' he yelled.

"'I'm fixing my radio,' I said, as I stuffed the computer in my book bag.

"'Do you always fix your radio in the john?'

"'No,' I said, 'but it's cold outside.'

"I made up a story about being a tourist on vacation, something harmless, like a literature graduate student. He asked for my driver's license and wrote down my name and number on a piece of paper.

"When he finished writing, he turned to me and said, 'Look, all I have to say is, don't fix your radio in the bathroom anymore, because it's not a good place to do it. A lot of times drunks come in here and pass out, and we have to check up on them. I just thought you might have been sleeping in there.' I'm sure the piece of paper with my name on it ended up in the wastebasket, but the experience really shook me up."

After their close call at Harrah's, Doyne and Alix the following day tried to play roulette at other casinos. But much of their time was devoted to troubleshooting the computer, with Doyne plugging in the soldering iron and working on the sidewalk at one of the half dozen gas stations in Reno with outdoor electrical outlets.

"Harry went bye-bye early in the trip, and Raymond never settled into serious play. The trip was one long technological failure. It was basically shadow boxing, with only two or three hours spent actually playing, but it helped me get the feel of the casinos, and I spent a lot of time talking to croupiers."

Traveling back across the mountains that night, Doyne and Alix reached Sacramento by three in the morning. At dawn Alix caught a plane for San Diego. Exhausted, Doyne pulled off the road and slept for an hour before driving back to Santa Cruz alone.

Letty flew up from Los Angeles to spend Christmas in Santa Cruz, and then she, Doyne, and Dan Browne packed themselves into the

Blue Bus for the Project's first trip to Las Vegas. Arriving by Greyhound from Silver City, Norman would meet them in Glitter Gulch. After the Reno trip, Doyne had worked nonstop on fine-tuning the system, and the Project was ready, he thought, for an assault on the world's gambling mecca.

Leaving Santa Cruz on New Year's Eve, they drove south to Paso Robles before swinging east to Bakersfield and the Sierra Nevada. "On a moonlit night with mist rising in the valleys, it was a beautiful trip," said Doyne. "Letty drove and I slept until Barstow, where we hit Route 15 and the main blast into Nevada. There was a steady stream of cars pouring out of Las Vegas on their way back to Los Angeles. For a hundred miles in front of us there was a ribbon of headlights coming over the hills, while off on the horizon at the end of the lights was an orange glow. You knew you were going somewhere, that something big was sitting out there in the desert."

Arriving at four in the morning on New Year's Day, they camped on a hill overlooking the lights of Las Vegas. Dan Browne woke later that morning to find a volleyball in the middle of the desert. Everyone considered it a good omen.

Las Vegas in the daylight is flattened and splayed on the horizon. Without the nighttime tunnels of light leading into it, the city no longer has any obvious entrance or center. It straggles over the prairie after which it was named (*las vegas* being Spanish for "the meadows") in a net of streets and buildings that look in winter as brown as the scurf of desert on which they sit.

They drove to the southern edge of town and found the neighborhood bordering the University of Nevada. Tom Ingerson, visiting his sister and brother-in-law for the holiday, had been promised a roulette demonstration. Doyne unpacked the wheel from the back of the Bus and set it up in the Zanes' living room. He resoldered a few wires, adjusted the condoms on his solenoids, and pronounced everything ready for the demonstration.

"Of course it messed up," he said. "It didn't work. I had come by that time to expect that whenever I tried to do a demonstration for someone, the computer would go on the blink. According to Murphy's law, demonstrations are more prone to equipment failure than playing sessions, because you're going to be embarrassed."

Over the next few days Doyne got enough equipment running to play a solo session at the Golden Gate Casino on Fremont Street.

By that time all the other Projectors except Dan Browne had left town, and the only computer working reliably was Raymond. Without a radio link, Doyne would again risk playing both data taker and bettor. One of the more cramped and disheveled of the downtown casinos, the Golden Gate caters to a clientele of truckers hunkered over the craps layouts and Sun City snowbirds catnapping in front of the keno board. Roulette is no big draw, and the casino's two tattered layouts are shoved hard against the wall. Limiting himself to betting dimes, Doyne hoped his incursion into enemy territory would do only statistical damage.

Wired with Raymond the computer, the batteries, solenoid plate, and toe clickers, Doyne walked into the Golden Gate for a try at beating his first Las Vegas wheel. Dan Browne stood near him at the table. Pretending to play roulette, Browne was actually compiling a success histogram for the computer's predictions. Working backwards from Doyne's bets to compare predicted and actual outcomes in the game, he would duck behind the slot machines every few minutes to record data.

"This was my first real session playing solo," Doyne said. "I wanted to lock into a wheel and compile enough statistics to prove we had an advantage. I was very nervous about the croupiers being suspicious of me. But I was intent on winning and showing that the computer worked. I was playing dimes, and I thought with stakes like that they couldn't get too upset with me."

The system did indeed work, with startling accuracy. Chips seemed to get sucked off the layout and deposited in front of Doyne with one winning bet after another. They accumulated into mounds that attracted the usual crowd of pit bosses and players sniffing the scent of Lady Luck. With a look of ferocious, almost idiotic concentration on his face, Doyne ignored everyone around him and concentrated solely on the microswitches in his shoes and the predictions being tattooed onto his stomach.

Within half an hour the usual problems set in — short circuits caused by wires pulling loose and sweat. "It could have been disastrous," said Browne, who was watching by then from behind the slot machines. "Doyne was doing fine, betting dime chips, when one of solenoid buzzers jammed and started heating up. He rushed off to the bathroom, fixed the solenoid, and returned. But not much later, it happened again. He had barely sat down to play when he had to jump up and exclaim, 'Boy, have I got the shits!'

"This happpened several more times, with the solenoid becoming progressively hotter, and Doyne running to the bathroom faster and faster, saying, 'I've got the shits today! Boy, do I have the shits!' On the last trip to the toilet the pit boss followed him and sat in a neighboring stall. The croupiers must have thought he was crazy, and I guess he was. Doyne was jumping around like a gnat on a skillet. But I'm sure they never dreamt of the poison that brought on craziness like this."

Doyne cashed in his chips and quit after four and a half hours in the casino. Betting dime stakes, in several hundred trials, the profit was small. But what mattered was the computer's advantage over the house, which he and Browne figured conservatively at 25 percent.

"I was elated," said Doyne. "We had proved we could go into a casino, set parameters, play on an unknown wheel, and beat the house by at least a twenty-five-percent advantage, which is a very hefty margin. Now all we needed to do was raise the stakes."

6

The Invention of the Wheel

The main thing is the play itself. I swear that greed for
money has nothing to do with it, although heaven knows
I am sorely in need of money.

Feodor Dostoyevsky

Although the paternity of the computer, the law of probability, and the game of roulette are all ascribed to Pascal, I can report — after extensive research into the matter — that while the first two offspring have an undeniable claim to his name, the third is illegitimate. There are, however, good reasons for implicating Pascal in the act of conception. This is a case of mistaken identity, but one made with good cause.

After inventing the mechanical adding machine, Pascal secured a monopoly on its exploitation and supervised the manufacture out of wood, ivory, ebony, and copper of more than fifty Pascalines. Among his contemporaries, his fame rested primarily on this invention (even the great Descartes applied for a demonstration). But Pascal increasingly turned his thoughts to more abstract speculation in gaming and theology, and on his way to an irascible sainthood at the Jansenist convent of Port-Royal, he paused briefly to invent the mathematics of probability.

In 1654 a friend had written him to ask if Pascal could solve the *problème des parties*, or problem of points. When players end a card game before its natural conclusion, how should they divvy up the pot? Pascal in turn posed the question to Pierre de Fermat, noted mathematician and jurist in Toulouse, and together, in a series of written exchanges, they worked out the mathematical basis for the theory of probability. In solving the problem of points, the stakes should be divided, said Pascal, according to the concept of mathematical expectation, that is, according to each player's probability of winning the game.

His contemporaries marveled that chance could have laws. Catching the tone of their stupefaction, Pascal said of his accomplishment: "Thus, bringing together the rigor of scientific demonstration and the uncertainty of chance, and *reconciling those things which are in appearance contrary to each other,* this art can derive its name from both and justly assume the astounding title of the Mathematics of Chance."

Gamblers and popular historians further ascribe to Pascal the invention of roulette. (The word comes from the French for "small wheel.") In early versions of the game, the Greeks spun shields on sword points, and the Roman emperor Augustus had a rotating chariot wheel installed in the gaming room of his palace. While these devices employed a wheel and a stationary pointer, the modern game of roulette utilizes a more complicated mechanism, with rotor and ball revolving in counterdistinction to each other.

The story about Pascal's inventing roulette most likely stems from a toothache that seized him one night at the Port-Royal monastery in the spring of 1657. According to the account of his sister, Pascal leapt out of bed and turned his thoughts to a particular mathematical problem. After several nights of pacing his room, he simultaneously cured his toothache and made his greatest contribution to pure mathematics. Pascal was searching for the formula to a curve known as the cycloid, or "Helen of geometry," because of the fascination it had exercised over the minds of scientists from Nicolas of Cusa to Galileo and Descartes. Pascal defined this curve as the line described by a nail on the rim of a rolling wheel as it rises from the ground and falls again to meet it. The problem "considers the rolling of wheels," he said, "and is for this reason called *roulette.*"

Whether Pascal actually experimented with rolling wheels remains a matter of conjecture, although he did in fact solve the equations necessary to describe the cycloid. Gottfried Leibnitz, on reading Pascal's treatise on the subject, *Histoire de la roulette,* ten years after its author's death, would generalize Pascal's equations into the integral calculus, which allows for the study of continuously changing quantities and remains to this day the great tool of modern mathematics. "Nothing astonished me so much," said

Leibnitz about the *History of Roulette*, "as the fact that Pascal seemed to have his eyes obscured by some evil fate; for I saw at a glance that the theorem was a most general one for any kind of a curve whatever."

Roulette — not the curve of the cycloid, but the game of chance we know today — has assumed its own important, if checkered, place in the history of mathematics and physics. Scientists interested in studying its laws of motion have included James and Daniel Bernouilli, Laplace, Poisson, Poincaré, Claude Shannon, and Edward Thorp, many of whom, on lining up at the baize to examine more closely the wheel's physical and statistical attributes, have discovered that the game in play often has less to do with the stately procession of planets and more with the greed of men.

Roulette makes its first official appearance in 1765 when Gabriel de Sartine, a police lieutenant who thought he saw in it a gambling device immune to cheating, introduced it into the casinos of Paris. During the Revolution of 1789 royalist émigrés fled with their roulette wheels to Bath and other British resorts. By the early nineteenth century the game had spread to Continental health spas at Wiesbaden, Bad Homburg, Baden-Baden, Saxon-les-Bains, and Spa itself.

After impoverishing the hapless émigrés and minor nobility of Germany, the casinos were closed by the Prussian government in 1872. One smart operator simply packed up his concession and transported it from Bad Homburg to the principality of Monaco. At the time of Louis Blanc's arrival, Monte Carlo, the capital of this independent nation, was described in a contemporary account as consisting of "two or three streets upon precipitous rocks; eight hundred wretches dying of hunger; a tumble-down castle, and a batallion of French troops." That the royal family and citizens of Monaco are considerably better off today is due to the fact that Blanc cut them in for 10 percent of the action, which soon became substantial indeed. With roulette as its main attraction, Monte Carlo ruled supreme as the world center of gambling up until the rise of Las Vegas after World War II.

Roulette first reached the United States, like many exports from France, via New Orleans. It was played extensively in our original casinos: the paddle-wheel steamers that plied the Mississippi with loads of cotton bales and confidence men. Roulette had come ashore to be indulged in privately or illegally in Saratoga, New York City, New Orleans, Chicago, and Denver by the time Las Vegas took the fateful step of legalizing it and other games of chance in 1931. The Fremont Street casinos offered, at best, a brawling sumptuousness, and it took another fifteen years before the entrepreneurial talent of a New York gangster named Benjamin "Bugsy" Siegel would turn Las Vegas into the gambling mecca of the world.

In 1946, just across the town line in a blighted stretch of greasewood known as the Strip, Bugsy Siegel opened the Flamingo Club, the first of Nevada's famous pleasure domes. Bugsy was gunned down by the mob within a year, but his idea had already taken hold. Siegel, in a stroke of genius, had seen that a completely encapsulated fortress of delight, designed around a theme of historic excess, such as that of Arabian sheiks, Spanish grandees, or Roman emperors, would become the ideal destination for late-twentieth-century tourism. Packaged into segments of one, three, five, seven, or more days, the nobility of leisure could be made available to everyone. Suitably elevated by the grandeur of Bugsy's ambiance and décor, the tourist could adopt other attributes of the nobility — their love of play, largeness of gesture, and disdain for monetary loss. What better environment, thought Siegel, for fleecing people at the tables?

Within a decade a dozen more pleasure domes had sprung up on the Strip. Today Las Vegas offers more than a hundred. France legalized gambling two years after Nevada, and roulette quickly appeared in casinos scattered along the coasts at Deauville, Biarritz, Nice, and Le Touquet. It could also be found at Estoril in Portugal, in Rome, Venice, San Remo, and Salzburg. When Britain legalized gambling in 1960, roulette was reintroduced as a strong component in what quickly became the country's leading industry. (Britons spend more than $5 billion annually on gambling, while Americans wager in excess of $100 billion a year.) Roulette today can be played throughout the Caribbean and South America, in Marrakech, Macao, Quintandinha, Constantsa, and even Mbabane, Swaziland, where a combination casino—hot springs offers the rare opportunity to see South African racists rubbing shoulders with Swazi tribesmen.

*

For all its regularity, the game of roulette in play produces an outcome at once random and unrepeatable, thereby perfectly illustrating the laws of chance. A roulette wheel consists of a precisely machined rotor that is balanced and spun on a steel spindle. Spaced evenly around the rim of the rotor are thirty-eight pockets numbered from 00 to 36. These alternate high-low, odd-even, and red-black, except for two green pockets, numbered 0 and 00, which face each other across the rotor. These green cups — and the fact that the payoff in roulette is less than thirty-eight to one — guarantee the house its advantage in what would otherwise be a fair game.

The roulette wheel presides at the far end of a long rectangular table covered in green baize and marked with the squares and columns of a betting layout. Arranged sequentially in three columns that increment from left to right and top to bottom are the thirty-six numbers that lie scattered in more random fashion on the wheel. At the head of these columns preside squares for the 0 and 00, while at the foot and along the near side of the layout lie other squares representing bets on red or black, odd or even, the first twelve numbers, entire columns of numbers, and so on. Because bets can be split between numbers, columns, and rows, a roulette layout offers the discriminating investor thirteen different types of wagers.

In the middle of a game, a roulette layout looks something like a Mondrian painting reworked by Jackson Pollock: dribbles of yellow, red, blue, and gold chips cover the white boxes etched on the baize and crisscross the lines between the boxes. Players stack chips into Pisan towers, change their minds, redeploy them from one box to another, and then, at the last minute, scatter a final inspiration of chips onto the green felt. Their work done, they stand back to watch its public reception.

The odds, at least in American roulette, heavily favor the house. A winning bet straight up on one of the thirty-eight numbers (including the 0 and 00) is paid off at the rate of thirty-five to one. If roulette were a fair game, with no house advantage, the bet would pay thirty-seven to one, not counting the chip originally wagered. Every bet on the layout is similiarly discounted, so that the house clears a cool 5.26 percent profit, except for the five-number bet that straddles the 0, 00, 1, 2, and 3, where the casino pockets a neat 7.89

percent profit. The odds in roulette compare unfavorably to blackjack, craps, and baccarat. In the last of these, for instance, the house advantage is 1.25 percent.

It matters a great deal to the mathematics of the game that roulette wheels in Europe have thirty-seven pockets, rather than thirty-eight, the house having eliminated the 00. It was originally removed by Louis Blanc when he took over the gambling concession at Bad Homburg in 1840, and his promotional device has since remained a permanent feature on the European wheel. This, and other differences in how the game is played, narrow the house advantage to 1.35 percent, which explains why roulette is still the pre-eminent diversion in European casinos and others outside the United States.

As John Scarne, magician and gambler-in-residence to the Hiltons, said of roulette, "This is the game which the handsome hero and well-dressed heroine have played for years in countless motion pictures, books and short stories. It has been publicized as the game of millionaire playboys, of kings and princes. It is celebrated in many stories of fortunes won and lost, of mathematical wizards who have spent years developing Roulette systems, and even in a song: 'The Man Who Broke the Bank at Monte Carlo.' Roulette is the world's oldest banking game still in operation, and through the years it has given rise to many true stories, as well as much that is legend and myth."

Given the rules of the game, there are three conceivable systems for beating roulette. The first is mathematical: a pattern of numbers, or procedure for placing successive bets, that would give the player an advantage over the house. Another kind of system relies on biased wheels and their tendency to favor one number over another. A third approach, through measurement of its physical forces, tries to predict the actual outcome of the game.

Roulette lends itself to the scrutiny of *systemiers*. Huddled at the wheel, scribbling figures into notebooks, these players resemble cabalists mooning over the number six. Gambling shops in Monte Carlo sell lists of roulette numbers recorded the previous day, and subscribers to the *Revue Scientifique* can receive their numbers monthly. Hundreds of books and articles describe "winning" schemes for mathematical prediction, but it remains indisputably true, after more than two hundred years of continuous play, that no such winning system exists. Edward Thorp is even more categorical in stating that "there is no 'mathematical' winning system for roulette and it is impossible ever to discover one."

Most of these systems employ a procedure known as "doubling up," which is based on the idea that a loss on a one-to-one bet can be offset by doubling the wager in each successive round. If you bet one dollar and lose, and then bet two dollars and win, you will have ventured three dollars in bets and "earned" four dollars in return, thereby making a profit of one dollar. As simple as it sounds, this system has two flaws, the first being that it requires an infinite bank. It may be improbable that one would lose nineteen times in a row, but to double up on an original wager of one dollar would require, in this instance, a bet of $524,288 in order to make the expected one-dollar profit.

The casinos, possessed of large but by no means infinite banks, protect themselves against doubling up with a simple countermeasure. They impose a house limit on bets, usually a thousand dollars, which effectively torpedoes the system at whatever wager would exceed the house limit. "What is perhaps truly amazing," says Thorp, "is that this is also true for all mathematical systems, no matter how complex, including all those that can ever be discovered," of which there are an infinite number!

Systems for doubling up, doubling down, tripling up, and so on, make up a class of strategies known as martingales. The word comes from the French expression *porter les chausses à la martingale*, which means "to wear one's pants like the natives of Martigue," a village in Provence where trousers are fastened at the rear. The expression implies that this style of dress and method of betting are equally ridiculous.

Another popular mathematical system is named after Jean le Rond d'Alembert, co-editor with Diderot of the *Encyclopédie*. The d'Alembert system, also known as the gambler's fallacy, operates according to the "maturity of chances" or "law of equilibrium." These "laws" maintain that a long string of numbers in one color increases the likelihood of the other color appearing in order to "average things out." This idea unfortunately contradicts the theory of probability, which asserts that every chance event is independent of the preceding and following events. Roulette balls have no memory, and their chance of landing on either red or black always remains an invariable fifty-fifty.

Roulette systems based on the discovery of biased wheels have proved more fruitful. A British engineer named William Jaggers once hired six clerks to record a month's worth of winning numbers at Monte Carlo. After calculating variances in their frequency, he and his staff cleared a tidy one million five hundred thousand francs by betting on the most favored numbers. They were stymied only when the casino redesigned its wheels with movable rather than stationary partitions between the cups. Switching these in the early hours of the morning, the croupiers were able to redistribute the variables on which Jaggers's system had relied.

Albert Hibbs and Roy Walford, friends from Cal Tech and fellow graduate students at the University of Chicago, managed a feat similar to Jaggers's in 1947. Using a Poisson distribution to distinguish biased from unbiased wheels, they found that more than a quarter of the wheels in Nevada were sufficiently unbalanced to overcome the house advantage. In a well-publicized session that spawned a rash of imitators, Hibbs and Walford cleared seven thousand dollars at the Palace and Harold's Club in Reno. Another graduate student at UC Berkeley by the name of Allan Wilson must hold the world's record for gathering statistics on biased wheels. He made not a penny for his efforts, but he and a friend, working in twenty-four-hour shifts manned over a five-week period, recorded no less than eighty thousand continuous plays.

Another Quixote in search of a system for beating roulette was the English mathematician Karl Pearson. Pearson invented the field of statistics. To him we owe such ideas as the normal curve, standard deviation, and the correlation coefficient — which, unfortunately, he used to "prove" that Jews are inferior to northern Europeans. Pearson took two weeks of data from the *permanences*, or daily run of numbers recorded in *Le Monaco*, and analyzed them for statistical fluctuations.

Publishing his findings in an article entitled "Science and Monte Carlo," Pearson wrote: "If Monte Carlo roulette had gone on since the beginning of geological time on this earth, we should not have expected such an occurrence as this fortnight's play to have occurred *once* on the supposition that the game is one of chance. . . . Monte Carlo roulette, if judged by returns which are published apparently with the sanction of the *Société*, is, if the laws of chance rule, from the standpoint of exact science the most prodigious miracle in the nineteenth century." Pearson's data had been faked by journalists preferring to keep their tally in the casino bar. But the scandal gave him the opportunity to call for closing the casinos and using their resources to endow "a laboratory of orthodox probability" designed to further Pearson's social Darwinism.

Feodor Dostoyevsky, given to epileptic seizures that left him in a state of near-idiocy for days on end, employed a roulette system that operated more in the realm of psychology than of statistics; it had to do with continence and emotional equanimity. "I do know the secret," he wrote to his sister-in-law after a gambling session at Wiesbaden, "and it is extremely stupid and simple: it consists in controlling one's-self the whole time, and never getting excited at any phase of the game. That is all; in that way one can't possibly lose and *must* win."

The difficulty with his system, as Dostoyevsky realized, is that "I have an evil and exaggeratedly passionate nature. In all things I go to the uttermost extreme; my life long I have never been acquainted with moderation." Sigmund Freud thought the system had other problems, of a sexual nature. "The 'vice' of masturbation is replaced by the addiction to gambling; and the emphasis laid upon the passionate activity of the hands betrays this derivation," said Freud in an essay titled "Dostoevsky and Parricide." "The irresistible nature of the temptation, the solemn resolutions, which are nevertheless invariably broken, never to do it again, the stupefying pleasure and the bad conscience which tells the subject that he is ruining himself (committing suicide) — all these elements remain unaltered in the process of substitution."

One of the more inventive roulette schemes is described by Alexander Woollcott in his story "Rien ne va Plus." With the success of Louis Blanc's operation at Monte Carlo, the casino owners in Nice and elsewhere on the coast tried to put a spoke in his wheel by broadcasting exaggerated reports on the suicide rate at Monaco. It was getting hard to swim there, they said, with so many dead bodies washed up on the beaches. Dining one night with friends on a terrace in Monte Carlo, the narrator of Woollcott's story is "eating a soufflé and talking about suicide." Earlier in the day he had watched a well-dressed young man lose all his money in the *salles privées* of the casino, and now there are reports that the man has been found dead on the beach with a bloodied shirt and a gun in his hand. To avoid publicity, agents from the casino tuck ten thousand francs into the corpse's dinner jacket pocket, "so that the victim would seem to have ended it all from *Weltschmerz*." But as soon as the agents are out of sight, the corpse jumps up. With tomato sauce still smeared on his shirt front, he races to the casino and uses their ten thousand francs to win a hundred thousand more.

Perhaps the most successful of all *systemiers* was Marcel Duchamp. Having already launched his artistic career into the outer reaches of Dada and surrealism, Duchamp in 1924 perfected a system for playing roulette in which "one neither wins nor loses." Forming a company to exploit his scheme at Monte Carlo, he designed an issue of thirty stock certificates for sale at five hundred francs apiece. These bear on their face a Monte Carlo roulette layout and wheel with a superimposed photograph of Duchamp taken by his friend Man Ray. The photo shows a satyrlike Duchamp with horns and a beard made out of shaving cream. Signed by Rrose Selavy *(eros c'est la vie)*, Chairman of the Board, these pieces of paper are now worth far more than five hundred francs apiece. Although Duchamp managed at the time to sell only two of them, this capital financed a month-long-trip to the roulette tables at Monte Carlo, where, to his great satisfaction, Duchamp broke even.

Given the impossibility of devising a winning mathematical system, the only feasible approach to beating roulette lies in *physical* prediction. This requires a device intelligent enough to comprehend its laws of motion and swift enough to calculate its outcome while the game is in play. It requires, in other words, a computer. Analog computers could be built to these specifications in the 1960s. Digital computers followed in the 1970s. Whether the idea preceded the technology, or vice versa, microcomputers and roulette prediction were lovers attracted to each other at first sight.

It was Edward Thorp who pioneered the use of analog computers to beat roulette. As early as 1962, in the first edition of *Beat the Dealer*, he declared himself in possession of "a method for beating roulette wheels whether or not they are defective!" In the gung-ho prose of a scientist trying for suspense, Thorp added cryptically: "I played roulette on a regulation roulette wheel in the basement lab of a world-famous scientist. We used the method and steadily averaged 44 per cent profit. In an hour's run, betting no more than $25 per number, we won a fictional $8,000!"

As to why he was writing a book about his success, rather than retiring to Cap d'Antibes, Thorp explained that "there are certain electronic problems which have so far kept the method from being used on a large scale in the casinos."

Seven years after mentioning it in *Beat the Dealer*, Thorp was mathematically more explicit, although still vague about the practical details, when he published his article in the journal of the International Statistical Institute. This is where he discusses briefly, in remarks directed toward readers with a background in probability theory, his development of the Newtonian and quantum methods for physical prediction. His collaborator, the "world-famous scientist," remains unnamed.

Only recently, in a series of articles written for the magazine *Gambling Times*, has Thorp been more forthcoming about the nature of his roulette project. The name of his partner also appears in print for the first time — Claude Shannon. While still a graduate student, Shannon had worked out the equations for comprehending switching electrical networks. The general expression of his findings, which became known as information theory, is now applied to switching networks as diverse as telephone exchanges, the computer, and the human brain. Shannon's "basic idea," as he put it, "is that information can be treated very much like a physical quantity such as mass or energy."

Shannon was in his early forties, a professor at MIT and a recognized luminary in applied mathematics, when Edward Thorp in December 1960 had the temerity to knock on his office door. Thorp had just finished his doctorate in mathematics at UCLA and taken his first job as an instructor at MIT. His dissertation was titled "Compact Linear Operators in Normal Spaces," but what really interested Thorp was the theory of gambling. Writing programs to simulate play in blackjack, he had a field day on the computers at MIT while devising both the basic and more sophisticated versions of his card-counting strategy.

About to announce his findings at a meeting of the American Mathematical Society, Thorp thought he should rush his talk into print and save his ideas from being pirated. He targeted the *Proceedings of the National Academy of Sciences*, a prestigious journal that publishes articles only on the recommendation of its members. The sole mathematician and National Academy member at MIT was Claude Shannon.

"I was able to arrange a short appointment one chilly December afternoon," said Thorp. "But the secretary warned me that Shannon was only going to be in for a few minutes, not to expect more, and that he didn't spend time on subjects (or people) that didn't interest him (enlightened self-interest, I thought to myself).

"Feeling both awed and lucky, I arrived at Shannon's office for my appointment. He was a thinnish, alert man of middle height and build, somewhat sharp featured. His eyes had a genial crinkle and the brows suggested his puckish incisive humor. I told the blackjack story briefly and showed him my paper."

Shannon quizzed Thorp for possible errors in his analysis. Finding none, he told him to condense the paper and change its title from "Fortune's Formula: The Game of Blackjack" to something more academically neutral. (It was published as "A Favorable Strategy for Twenty-One.") Shannon then asked if he was working on other problems in gambling.

"I decided to spill my other big secret," said Thorp, "and told him about roulette. Several exciting hours later, as the winter sky turned dusky, we finally broke off with plans to meet again on the roulette project."

Shannon lived in a large wood-frame house on one of the Mystic Lakes north of Cambridge. Down in the basement he and Thorp went to work in a laboratory that Thorp described as "a gadgeteer's paradise. It had perhaps a hundred thousand dollars worth of electronic, electrical and mechanical items. There were hundreds of categories, like motors, transistors, switches, pulleys, tools, condensers, transformers, and so on." The two men ordered a regulation roulette wheel from Reno and set it up on Shannon's billiard table. Around the wheel they stationed a strobe light, a clock, a movie camera, and the switches needed to coordinate the strobe and clock while filming the ball in motion. Their research — which agreed with that later conducted by Eudaemonic Enterprises — concluded that roulette is a highly predictable game.

Thorp and Shannon then set to work to build a computer. They came up with a transistorized analog device the size of a cigarette pack. It received data via four push buttons, which were compressed on successive revolutions of the rotor and ball in front of a fixed point. Because theirs was an analog computer, representing variables by means of voltages, Thorp and Shannon were limited in the complexity of their program, and they either ignored or were ignorant of many factors needed for physical prediction on any but tilted roulette wheels.

Claude and Betty Shannon, Edward and Vivian Thorp, and their computer checked into the Riviera Hotel on the Las Vegas Strip in 1962. Thorp was already well known in Nevada, having made a big splash the previous year publicizing his card-counting strategy. Two professional gamblers had staked him $10,000 to prove his system, and he had parlayed that sum, during his spring vacation from teaching at MIT, into $21,000 — a tidy profit of more than 100 percent. Having thus launched a million card counters as a plague on their tables, Thorp was not a welcome sight to the casino owners of Nevada. Later he was barred from play and took to wearing disguises; he grew a beard, donned wraparound sunglasses, and traveled always in the company of friends. But at this stage of his notoriety he still had access to most of the clubs, and none of them suspected that he was out to beat them at a game other than blackjack.

The Thorps and Shannons spent a week at the Riviera. They took in some shows, lounged around the pool, played diversionary blackjack, and did their best to beat roulette. For security reasons they had developed a two-person system, with a radio transmitter built into their "cigarette pack" computer. The radio informed the bettor of the winning octant by means of a do-re-mi scale whose tempo was calibrated to the ball-wheel configuration it was meant to be mimicking. These radio signals were picked up by a hearing aid and "a little bitty loudspeaker with flesh-colored wire attached to it, which we shoved into our ear canal. The trouble," said Thorp, "is that the wires kept breaking. So we shopped around and got steel wires about the size of a hair, but even these were fairly fragile.

"Sometimes I was the bettor, and sometimes I was the data taker. We traded off. But it would take us a while to get wired up in our hotel room and get in there. It was a real hassle putting it all together. It's a long, tedious project, even though it's conceptually very simple."

"Difficulty with read-out devices" is Thorp's modest confession in print as to why he and Shannon gave up on their roulette computer after three or four sessions. During a recent conversation in his office at the University of California at Irvine, Thorp was more graphic in describing a general snafu of broken wires, gibbering beeps, shocks, and other electronic failures. He and Shannon tried sporadically over a number of years to debug their system, until they finally abandoned it as a bright idea whose implementation had eluded them.

Their computer ended up in Shannon's basement, "where it's gathering dust," said Thorp. "If I were doing the whole thing over today, I'd use digital technology, a microprocessor. It's really the way to go. You don't have to use linear approximations or the other kinds of approximations that analog computers like. You could solve the equations and put in the right curves." He stopped and sat blinking behind his desk, as if frightened by a moment of misplaced enthusiasm. "But I'd never go back to it," he said quickly. "It would require a huge amount of work."

It was the latest advance in computer technology — from analog to digital microcircuitry — that provided the breakthrough chance for roulette prediction. Understanding their different modes of operation explains why, in this instance, digital computers are superior to analog. Analog computers, so named because they work through electrical analogs, represent variables by means of voltages. Numbers are correlated, like those on a speedometer, to continuous gradations in physical quantity.

"To make an analog roulette computer," as Norman explained it, "you build a circuit which mimics electronically what's happening physically to the ball and rotor. You model these forces by means of a voltage that decreases with time, but you make the model operate ten times faster than the game in play.

"It's simple enough to program an analog roulette computer. But if you want to change your algorithm — the equation you're using to predict what's going to happen — you have to rewire the computer, because the program in an analog device is the circuit itself."

Rather than operating by means of electrical analogs, digital computers, as their name implies, function in the realm of digits or numbers. This allows them nearly infinite storage capacity, great logical facility, and easy programmability. Instead of having to solder in new transistors and rewire circuits, the modification of a program in a digital computer requires nothing more than the addition of a number.

"We opted for digital over analog because it was more flexible," Norman said. "It allowed us a wider range of predictive models without having to mess around with changing the circuitry for every model. It's also more general purpose. Not only can a digital computer make the predictions, it can also send out signals to transmitters, communicate with toe clickers, and generate solenoid buzzes."

Although technologically superior, there was one difficulty entailed in opting for a digital over an analog computer. It meant that the equations of motion involved in predicting roulette, rather than being *approximated* by continuous gradations of electrical analogs, had to be *solved*. Digital computers tolerate no approximations. They operate solely in the black and white world of numbers ordered into equations with solutions. Where Thorp had experimented with an empirical system based on what he thought were standard values taken from a normal roulette wheel, Eudaemonic Enterprises was striving after a system at once more universal and specific. They hoped to solve the equations governing *every* kind of behavior in the roulette cosmos. And at the same time they wanted algorithms flexible enough to account for the minute differences found with *each* specific wheel. Neither they nor their microprocessor were prepared to tolerate any behavior left unexplained.

7

Strange Attractors

The squirming facts exceed the squamous mind,
if one may say so. And yet relation appears . . .

Wallace Stevens
"Connoisseur of Chaos"

In the spring of 1978 Eudaemonic Enterprises, nearing its second anniversary as a company, announced that the Pie was about to be served. There would be more than enough to go around, but anyone hoping to be first in line for a taste of roulette richness was advised to show up in Santa Cruz immediately. From up and down the coast Eudaemons arrived and threw themselves into round-the-clock sessions building computer hardware, practicing on the eye-toe coordination machine, designing costumes, discussing gambling theory, and attending classes, complete with a twenty-five-page manual and homework, on how to beat roulette by computer. This frenzy of soldering, sewing, clicking, and betting was directed toward an imminent junket to South Lake Tahoe and Reno. Doyne's winning session that winter at the Golden Gate had already proved the efficacy of the system. Designed to be free of broken wires and other snafus, the computer was now ready, they thought, for its first major raid on the casinos.

With people camped in the house and out in the yard in sleeping bags, 707 Riverside looked like an ashram devoted to studying the tao of physics. Friends provided background music on the piano, while Ralph Abraham dropped by for an occasional chat on gambling theory and casino disguise.

"Ralph and I spent a lot of time," said Doyne, "talking about what it takes to convince the casinos you're real." They also discussed problems raised by Richard Epstein in his book *The Theory of Gambling and Statistical Logic*, which Ralph was teaching that spring in a course on the mathematics of gambling. Given a bank of x number of dollars, what percentage of it do you want to wager at each play of the game? Is it advantageous to bet on one or more than one number at a time? For a more detailed examination, these questions were referred to Alan Lewis, a specialist in statistical mechanics.

Lewis closeted himself in Doyne's bedroom with the roulette wheel and Raymond, where he gathered data for a couple of hours a day before hitting the beach in the afternoon to work on his tan. His task was to find out the computer's exact advantage over wheels tilted at different angles and spun at different speeds. A young instructor at the university with degrees from Cal Tech and Berkeley, Lewis tooled around town in a Triumph Spitfire and sported a polyester wardrobe that made him look more corporate than academic, which was appropriate considering that he later left the university to play the stock market. Laid back and laconic, he was amused by the Project and put in a lot of work on it, both theoretical and practical.

Charlene Peterson, a friend of Doyne and Letty's from Stanford, was also newly enlisted into the Project. Charlene had saved enough money working as a cocktail waitress at Ricky's Hyatt House to buy land in northern California, and now she and her boyfriend were accumulating their last stash of capital before dropping out for good. She was teaching primary school in the Santa Cruz Mountains when Doyne tapped her as a potential Eudaemon. He imagined that Charlene — "the beautiful Girl Scout type," with the patter of a cocktail waitress initiated into Zen Buddhism — could replace Alix in the role of high-stakes bettor.

John Loomis was another Stanford friend recently recruited. Along with talents as an art historian, Italian cook, and singer of Tuscan folk songs, Loomis was good with his hands. He was working as a carpenter in the Bay Area when Eudaemonic Enterprises brought him to Santa Cruz to perfect the input-output devices. After taking the condoms off the solenoids and discovering that rip-stop nylon worked even better to hold them in place, he then built a new apparatus for fitting the buzzers securely against the stomach.

Nerds — the term for goggle-eyed technicians whose dreams oscillate between wiring diagrams and *Playboy* pinups — were not employed by Eudaemonic Enterprises. But they did have a hacker. This is a more specific word describing someone possessed by the beauty of computers and their programs, a person so in tune with the new technology that it responds in his hands with wild inventiveness. Jim Crutchfield was proud to consider himself one of this species. Endemic to the Silicon Valley and other high-tech regions, hackers communicate by means of a language as distinct as Serbo-Croatian or Basque. It defines a close-knit community and distinguishes those who really know computers from those who merely use them. With nary a reference to the English language, a hacker in full throat can shape entire paragraphs around the syntax of *baud rates, warm boots, down-loaded programs, bits, buffers, babble*, and other binarisms.

A San Franciscan who wears his brown hair parted in the middle and long over his ears, Crutchfield surfs, snorkles, and backpacks. But what he really cares about in life are computers. The intense and abstracted look on his face, the compression along the forehead, the mumbling to himself, the diffidence, and an inability to chitchat in the small coin of everyday speech are signs of a hacker more interested in talking to machines than to humans. "The real distinction among computer users isn't between theoreticians and experimentalists," he said. "It's between hackers and nonhackers. The hackers are the ones who understand systems and how to use them."

As an exemplary tale of hacking — and warning about a way of life in danger of selling its soul to the highest bidder — Crutchfield told me about his days in the Home Brew Computer Club. Long before talk about "computer literacy," a group of nerds, hackers, freaks, and college students in Silicon Valley used to get together and compare notes. "One night," said Crutchfield, "two long-haired hackers showed up with a PC board connected to a screen capable of doing primitive graphics. Dope-smoking hippies like the rest of us, they had wired the thing together in their garage — except that this was Stephen Wozniak and Steven Jobs, and they happened to be carrying the first Apple computer.

"All of us knew we were looking at something important, but I'm amazed at how fast the knowledge has spread out into the general population. As Xerox and IBM move in to package everything, I wonder if there's still going to be room for the hackers. That's where this country is still way ahead of the Japanese. They have such a highly structured society that they don't have any place in it for misfit hackers. But these are the people coming up with the truly creative ideas. This is where you get the real advances."

Besides lending his expertise to Eudaemonic Enterprises, Crutchfield was also hacker-in-residence to a special research group at the university that included Norman Packard and Doyne Farmer. Doyne had been lured back to graduate school to work on the physics of chaos. New theories were in the air for describing chaos, or at least some of its simpler manifestations, by means of geometrical structures known as strange attractors. Some of the earliest research into these structures — news of which would soon shake up the entire profession — was being done at Santa Cruz. When not examining chaos in their laboratory at the university, the Projectors were playing roulette down on Riverside Street, and for several years these two inquiries were pursued with equal passion.

The study of strange attractors has recently become the hot topic in nonlinear dynamics, which encompasses behavior that used to be dismissed as turbulent or random or otherwise too complicated to be explained. To simplify their lives, physicists from Galileo to the present have theorized about stable, linear systems. These are exceedingly rare, however, and what real life mostly offers are non-linear, unstable, dynamic, and chaotic systems — things like cloud patterns and flowing water and the firing of synapses in the human brain. Contemporary physics is currently in the midst of a revolution as it turns from the study of linear to nonlinear systems. But it was only a few years ago that the entire field, with no more than a handful of researchers, was wide open to scientists accomplishing the breakthrough work in isolating strange attractors and defining some of the simpler forms of chaos.

Calling themselves either the Dynamical Systems Collective or the Chaos Cabal, Crutchfield, Packard, Farmer, and their fourth member, the young physicist Robert Shaw, would make a name for themselves doing a good bit of this seminal research. But when they first got started, they had nothing going for them other than a combination of talent and luck. And computers. The new physics is too complex to be done without them, and the Chaos Cabal — all of them by then hackers in their own right — happened to be more knowledgeable about computers than any other physicists around. Their hours spent in the Project Room programming, soldering, and debugging roulette computers would pay off in unexpected ways. They had reached the point where they could patch together a system and write a program to study the physics of anything, from roulette balls to strange attractors.

The KIM and the roulette computers were fine for solving Newtonian and quantum equations. But chaos is a tougher nut. Cracking it requires a computer big enough for what hackers call number crunching. Poking around in the basement of the physics building one day, the Chaos Cabal found an old Systron Donner analog computer left over from an engineering department that never got built at Santa Cruz. Old and dusty, the machine was also monstrous in capacity; so they dragged it into an empty office and got it up and running. Adding a collection of Z-80 based micros and a NOVA digital computer cadged from the experimental high-energy physics group, the Dynamical Systems Collective wired itself into a rat's nest of plotters, printers, terminals, and screens. In 1979, less than two years after hanging its name on the door, the Chaos Cabal was reputable enough to get its first National Science Foundation grant.

While still an undergraduate at UC Santa Cruz, Crutchfield had done much of the work required to get the analog computer running. He then built an ingenious interface that allowed it to talk to the digital computers, although they don't speak the same language at all. On finishing college he started publishing scientific papers as a full-fledged member of the Chaos Cabal, which, oddly enough, caused the university some embarrassment. Having no use for hierarchies or status, Crutchfield had resisted going to graduate school. But here he was wandering in and out of the physics department, doing research in its laboratories, and publishing papers exactly as if he *were* a graduate student. Arriving at a de facto compromise, Crutchfield allowed his name to be put on all the forms that would make him an official student.

In the spring of 1978, like the rest of the Chaos Cabal, he diverted his attention from strange attractors to roulette. Crutchfield cut himself in for a slice of the Pie and moved down to Riverside Street to become Eudaemonic Enterprises' hacker-in-residence. Generally in charge of perfecting the equipment, he also built a third roulette computer. Joining Raymond, the prototype, and Harry, the first of its offspring, there was now a new member of the Eudaemonic computer family named Patrick, after James Patrick Crutchfield.

"Raymond at this point had already been put out to pasture," said Doyne. "So for playing in the casinos we had Harry and Patrick. Jim made connectors for them, nicely painted and color-coded. He built little aluminum boxes about the size of an address book to hold the computers, and when he had finished going over them, the hardware and peripherals were generally of a much higher quality."

As computers, Harry and Patrick were completely self-contained, with the memory and logical ability to play roulette on any wheel they chose. All they needed for this assignment were peripheral interface devices, which for the Project consisted of toe-operated microswitches for getting information into the computer and thumping solenoids for getting it back out. Because Eudaemonic Enterprises had opted for a two-person system, each roulette team would have to be outfitted with not one but two computers, and these in turn would need a way of communicating with each other. This was being handled by magnetic induction, which used a wireless transmitter and receiver that can be spoken of, without splitting hairs, as a kind of radio link. To simplify the discussion further, one could characterize the Project as having engineered a two-person system that employed both a transmitting and a receiving computer. These were virtually identical in size, and both possessed a full complement of solenoids, although only the transmitting computer was programmed with a roulette algorithm and endowed with toe-operated microswitches.

The first of the Project's receiving computers had been built by Ingrid Hoermann as her homework assignment in Physics 107, the basic electronics course for physics majors in which Doyne was her teaching assistant and Norman her tutor. The computer was christened, according to Eudaemonic tradition, with Ingrid's middle name, Renata. A second receiving computer was built that spring by Norman. As the computer Harry already bore his middle name, and as the Project — following the customary nomenclature — considered their receiving computers female in gender, the new machine was named after Norman's youngest sister, Cynthia. Descended from grandmother KIM and father Raymond, the Eudaemonic family now included four sleek little computers — the two transmitters, Harry and Patrick, and the two receivers, Renata and Cynthia.

A classical pianist majoring in music, Ingrid had been about to graduate from UC Santa Cruz when she decided in the fall of 1976 that she "wanted some sort of general education, and playing the piano wasn't giving it to me. I spent entire days working on jumps and octave leaps. But it was just physical stuff, without any kind of intellectual discipline.

"I always end up going overboard, being more thorough than I need to be, covering something too far and then moving back to where I wanted to be in the first place. So after finishing all the requirements needed for a degree in music, I put in a request to stay at Santa Cruz another year or two. I told them I wanted to be a sound engineer and study physics."

Ingrid worked as a recording technician for the music department. She got involved in electronic music performances and other "happenings." She built a small synthesizer and learned a lot about the physics of music. It was this interest in synthesizers and contemporary musical performance that had steered her originally into working toward a degree in physics. To help her through the science curriculum, Doyne introduced Ingrid to Norman, and the two of them arranged to swap instruction in physics for piano lessons. Doyne also set her up with a nifty class project bound to get her high marks if she could pull it off, and by the end of the quarter Ingrid had succeeded in building her first computer.

"Ingrid learned more about microcomputers that quarter," said Norman, "than she probably cares to admit, and she got a very good evaluation for her project."

"Norman is a very talented piano player," said Ingrid in her own evaluation. "But when it came to tutoring me in physics, we spent most of our time talking about either roulette or strange attractors."

A regular visitor to Riverside Street, Ingrid soon moved in as a full-fledged Projector. "I had always felt privileged when they invited me to stay for dinner," she said. "It would be a huge, family-style affair with all these crazy, high-powered, entertaining people. We would sit around the table and talk for hours. They were scientists, but none of them was a nerd. They read books and knew about other things.

"I had never been in a communal household that worked like this, where people really enjoyed each other. Everything in the house was supposed to be run by popular vote, and we had a real sense of responsibility toward the place. We shared *everything* — food, toothpaste, tools. Each person had a cooking night and was responsible for having dinner ready. There were rules on how things should go with recycling and shopping and planting the garden. We were supposed to talk about any problems at occasional house meetings held at dinner, although it was hard to find a meal without friends over. A job wheel with our names on it listed all the tasks that needed to get done. The wheel rotated once a week, so that everyone eventually cleaned the bathrooms and kitchen or mulched the compost or watered the garden. For the big jobs, like a massive cleaning once every three months or building the fence around the back yard, we assumed the whole house would help. We also hoped that visitors staying for a week or more would buy food and cook a meal. The point was to have an open house with enough room to welcome everyone."

Given to wearing Mexican sandals, blue jeans, homemade vests, and shirts open at the neck, Ingrid was at once playful, reticent, bold, and unpredictable. With dark hair and blue eyes, often glowering with concentration, she was not attractive in any way appreciated by a culture of coastal blondes. But she possessed another form of beauty having to do with energy and forcefulness of personality. She contained those qualities that the American Indian ascribes to Coyote. Mischievous and easy to smile, she could mimic other people's mannerisms with devastating accuracy. In her presence one was always somehow off balance, and therefore open to doing completely unexpected things. Resembling a geometric structure verging from order into chaos, Ingrid herself became one of the strange attractors around which the Project oriented its non-linear dynamics.

"When I learned about the Project," she said, "it was the best thing I had heard in a long time, sort of like a twentieth-century cowboy story. I was over for dinner at Riverside when Doyne, Norman, and I first went into the Project Room to look at the roulette wheel. We spun it and stood around and stared into it like a fireplace. I imagined what it would be like to go to Las Vegas and play the casinos, and it became incredibly glamorous.

"That's when I joined. I wanted go right away. It seemed like an alternate reality. You could lead this underground life and skip the eight-to-five world altogether."

Other than building Renata the computer, driving to the Silicon Valley for chips, compiling histograms with Alan Lewis, and learning her way around the mode map for a try at winning the Riviera Sweepstakes, Ingrid also designed some of the Project's costumes. "We needed clothes that would hide the computer and still be reasonably fashionable. After rummaging through people's closets looking for big sweaters and baggy blouses and dresses, we came up with some wonderfully sleazy combinations that Marianne, Charlene, and I then sewed into outfits."

Radio reception between data taker and bettor required that coils of antenna wire be carried somewhere on their bodies. The Projectors imagined sewing antenna cuffs onto their pants or antenna belts around their waists, but settled finally on a design for antenna T-shirts in which loops of wire were worn as a yoke around the shoulders. For concealing the computers, after much experimentation, they developed two sex-related systems. The men employed sacroiliac belts slung across the chest and worn like holsters for hidden weapons. One belt held the computer nestled under the left armpit; another held the batteries under the right. The women wore their computers and batteries snapped into leotards with pockets that fit under the breasts.

"The leotards were a pain to put on," Ingrid declared. "First of all, you had to take off all your clothes to get into them. They sagged under the weight of the computers and were so tight that our wires kept breaking. We later switched to bra and girdle combinations with hooks running up the front." Everyone, regardless of sex, also wore a solenoid plate with thumpers on the stomach.

Inspired like Ingrid by roulette madness, Marianne sewed her first three-piece suit. Specially designed for gambling, it had extra pockets in the vest and snaps under the arms for battery packs and computers. Naturally speedy, as if fresh from five cups of coffee,

Charlene topped even Marianne in high-energy output. She specialized in word games, puns, and other synaptic leaps, which kept the three of them amused as they rushed into the production of antenna T-shirts, computer sacroiliac belts, padded leotards, ripstop solenoid covers, and socks with holes hemmed into them.

"When it came to getting dressed to play roulette," said Ingrid, "there was a real ritual, like being a secret agent. To get everything on and hooked up took an hour. Then you conducted tests, and something invariably wasn't working. So you had to strip down all the way to your antennas. You tried to keep everything in discrete units, so there would be parts of the system and little piles of clothing scattered all over the room. People would be completely serious and upset about something not working, but they'd be standing there in their underwear clicking the computer with their toes. It was ridiculous.

"I had trouble finding a costume because the batteries were so bulky. I scrounged a pair of burgundy pants out of the basement and borrowed a top from Lorna — a floral print on rayon that crossed over in front with a big sash, like a Japanese kimono. I got dressed in the following order. First I put on a Playtex midriff bra that hooked in front. It had twenty hooks, and I invariably missed one in the middle; so I'd have to start hooking them all over again. The sacroiliac belts would have flattened me out completely, so for holding the computer and batteries, I stitched pockets onto the bra. Then I stuffed them with washcloths to fill them out. I figured if I were stacked I would look better, but mainly I needed to hide the corners. I also tried to keep the equipment away from my skin, because if I sweated I had a real problem with getting shocked, which was no fun at all. To prevent this, we later wrapped the computers in plastic bags. Then I put on the antenna T-shirt, which had connectors running down the front to the computer. Another set of connectors ran fron there down to the solenoid plate, which I wore on my stomach under a red and white polka dot girdle.

"I looked like a battle-ax on top. What made it even funnier was that I just didn't move in any natural way. The wraparound blouse was supposed to be flamboyant and sexy and loose, and I was wearing it loose, but I didn't want to reach over the table too far because someone might see my corners. I even took to ratting my hair, because I thought it would make my head look bigger and my body smaller. Thinking I'd try to fit in, I wore mascara and blusher and carried a handbag. When I was all dressed up, I was in an altered state of consciousness, and I was always cold, because when I'm nervous, even if I'm sweating, I get the chills."

"As you can imagine," said Doyne, "with so many roulette freaks at large in the house, the pandemonium was incredible. Norman was staying up all night working frantically on the receivers. At one point he said they were finished and took off for Portland to see Lorna, but I suspected they weren't really working. Jim Crutchfield was trying to debug Patrick and coming up with glitches all over the screen, until he later discovered one bad bit in the PROM. People were finishing end-of-the-quarter exams. It was crazy. Very crazy.

"Like always, we were way behind schedule. John Loomis had to pull out, and if we'd been smart, we would have scrapped the spring trip. But I felt as if we had built up so much expectation that we had to get out of town and do something different from living on top of all this stuff."

With Patrick still on the fritz from the New Year's trip to Las Vegas, and only Harry available for practice sessions, the Projectors decided to pack up the wheel and finish training themselves in Nevada. Riding in the Blue Bus — converted as it was into a mobile electronics workshop and casino — were Dan Browne, Charlene, Ingrid, and Doyne. Alan Lewis and his girlfriend, Molly, drove up in another car, while Marianne had gone ahead to meet Ralph Abraham at what he called his "cabin" — a three-bedroom condominium with Jacuzzi that he rented for the winter on the north shore of Lake Tahoe.

At the wheel of the Bus, speed-rapping with Dan Browne about cybernetics, Charlene circled twice around Hayward and got hopelessly lost outside Davis. Four hours later than expected, the Projectors reached Ralph's "cabin" in the middle of the night. "They rolled up and took over the place," said Marianne. "In no time they had the wheel set up and the house overrun with paraphernalia." They spent two days training themselves, and then late on the evening of the second day Doyne, Ingrid, and Marianne drove across the state line to Reno. With Patrick not yet debugged, and Norman's radio transmitters on the blink — which left the computers Renata and Cynthia unusable — Doyne would have to play solo with Harry.

Ingrid had never been inside a casino before, although she remembered when she was young that her mother on trips into the mountains used to stop in Tahoe to feed the slots. Ingrid would wait in the parking lot, and once her mother had returned to the car with a purse full of coins. Even though she had never seen the real thing, Ingrid knew the layout of a roulette table cold. Working from a cloth model painted by Alix for her own maiden trip to Reno, Ingrid and Marianne had developed lightning reflexes for covering the layout with chips. "To pick up the solenoid buzzes and get chips on four numbers at once, you had to be really fast," said Ingrid. "Even for three numbers you had to be quick. But we got good enough to do four numbers without any problem."

The Blue Bus pulled into Reno on a cool night in April. "We split up to check out casinos," said Ingrid. "We would meet later to compare notes on wheel tilt and other conditions for playing. Walking into my first casino — I don't remember which one — I was trying to act real casual, but I was so nervous I couldn't even get up the nerve to go over to the roulette tables. I was hanging around the slot machines, trying to watch from there, and the house cops must have thought I was underage and loitering. Two of them came over to ask for my ID, and I wasn't carrying one. So they threw me out for being underage. I felt terrible. If the first time I go into a casino I get kicked out, I figured I just wasn't making it in the Project."

When they regrouped later, Doyne reported finding a good wheel at Harold's, the biggest and flashiest of the downtown clubs. The three of them walked up to the roulette tables and Doyne bought into a game. He would play alone as data taker and bettor, while Marianne and Ingrid compiled statistics. "We did our best to look glamorous," said Ingrid. "Marianne was wearing makeup and trying to use her shoulders. I was dressed in a rabbit fur coat, and we were hanging on Doyne's arms like a pair of groupies. Doyne was nervous and staring intently at the wheel without being able to say a word. It wasn't obvious why anyone would hang on to that sort of guy. We stood around trying to look casual, and then Marianne and I would giggle and run off to the bathroom to pull out our notebooks and write down data." At four in the morning, down

a few dollars and getting shocked by a flaky computer, Doyne
called off the session. They drove back to Tahoe and reached the
north shore of the Lake at dawn.

Stripping off antenna T-shirts, sacroiliac belts, solenoid plates,
pimp shoes, midriff bras, wires, and switches, they jumped into the
hot tub outside Ralph's condominium. They watched the sun come
up and turned pink in the water and pinker still when they got out
to roll around in the snow. Later in the morning, after loading the
wheel and computers into the Blue Bus, they made their way over
Donner Pass and down the Sierra Nevada to Davis, where they
stopped for a pizza. Inside the restaurant, after staring into his
water glass for a long time, Doyne reached into his pockets and
pulled out all his money and keys. Handing them to Ingrid, he said,
"You take care of these. From now on, you make all the decisions.
I'm not taking any more responsibility for the rest of the trip."

"Up to that point," he said, "I had been trip leader and camp
counselor. I had been coping with everything from sewing holes in
my socks to rewiring computers. I was totally fried."

8

Exploring the Envelope

Lest men suspect your tale untrue
Keep probability in view.

John Gay

Alan Lewis was teaching a graduate course in electricity and magnetism when he and Norman, chatting after class one day, first talked about roulette. Of medium height, with brown hair and eyes, Lewis is given to understatement. Comprehending his jokes demands a minimalist aesthetic. He speaks in a monotone so precise that one sometimes feels the urge to record his voice and play it back at faster speed. But Lewis knows how to relax in a world in which there are more days you can drive your car to the beach with the top down than not. Born in Tucson and brought up in various parts of California, he is a westerner of the laid-back and unassuming sort.

He is also an expert in statistical mechanics. This is the branch of physics that attempts through mathematical models to analyze motion and the forces that cause it. Developed in the late nineteenth century by James Clerk Maxwell, Ludwig Boltzmann, and Josiah Gibbs, this statistical approach to physics — which originally described the atomic behavior of gases — has lately found new applications in gambling theory. As far as Lewis is concerned, the most interesting area of research for statistical mechanics is the stock market.

"In playing the stock market," he explains, "the trick is to take the gambling idea seriously. One is then led to a wealth of ideas from the well-developed mathematical disciplines of probability theory and statistics." With his theories on how to model the stock market like any other physical system, Lewis eventually left the university and moved to Newport Beach, California, as a stock market analyst. It is not your average stock broker who is interested in employing a physicist, and Edward Thorp once again reappears as a figure in this story.

In the fall of 1978 Lewis went to work for a small company called Analytic Investments, which manages $500 million for customers such as New England Telephone and Yale University. Located in one of the suburban haciendas strung along the freeways of Orange County, AI occupies a modest seven-room suite decorated with seascapes and wandering Jews maintained by a plant rental service. The company is run by Sheen Kassouf, a courtly New Yorker of Lebanese descent who specializes in playing the options market. With a degree in economics, Kassouf had hung out in New York for many years as a self-professed "boardroom bum" who spent his days camped in front of ticker tapes while looking for the American El Dorado: a stock market system that would allow him to win whether the market went up or down, on bear and bull days alike. He finally found what looked like a winning strategy through the use of bets known in stock market parlance as hedges.

Thinking along similar lines as Kassouf, and arriving at the same conclusion, was Edward Thorp. Then a professor of mathematics at UC Irvine, Thorp had moved on from blackjack and roulette to look at schemes for playing the stock market, which he viewed as another favorable game whose odds, under certain conditions, run strongly in favor of the informed bettor. Thorp and Kassouf met in 1965 when the latter was being interviewed for a job at Irvine. Sharing their ideas, they worked up a winning stock market system based on warrants and other kinds of hedges.

"I got interested in taking my gambling winnings and going into the stock market," said Thorp, "because it seemed like a larger-scale gambling game with a lot of extra interest to it and a lot more potential. I wouldn't run into the problems I had in Nevada with blackjack cheating and that sort of thing. If there was cheating, it wouldn't be directed at me personally, but at everybody who happened to be playing. So I did a lot of reading, and in the summer of 1965 I came across the idea of warrant hedging. I met Sheen as a prospective faculty member and found out almost immediately that he had the same idea, and had actually been doing some of this for a few years in a fairly crude but effective way. So we put our heads together and tried to improve the product, which developed into *Beat the Market.*"

Written as a sequel to Thorp's first book, the work he co-authored with Kassouf is titled, in full, *Beat the Market: A Scientific Stock Market System.* As Thorp put it, "Our general styles of doing things are somewhat different," and one senses this disparity even in the title of their book. The soft-spoken Kassouf contributed the "science" and "system" to the subtitle. But when it comes to "beating" — whether it be dealers, markets, or other running dogs of the capitalist system — Thorp's your man. He walks on his toes with the nervous lope of a street fighter looking for action. Others have seen in him a resemblance to Clark Kent. Remove the glasses and briefcase, and out of his office could emerge a muscleman from the beaches of Santa Monica, or a high roller just in from Vegas. Tanned and nervous, clothed in short-sleeved shirts and a full suit of what Wilhelm Reich called character armor, this eminent academician and winning gambler knows as much about tenure committees as he does about casino mobsters, and he probably thinks of them in similar terms. Thorp has never adopted the intellectual's disdain for worldliness. He wants to be rich and famous, and he makes no bones about it.

After their book appeared, Thorp and Kassouf split up to incorporate themselves into companies. Kassouf went public, while Thorp the high roller looked to private individuals for capital. He started a hedge fund called Princeton Newport Partners, named after the two cities in which its principal offices are located. To avoid governmental regulations, Princeton Newport works with unlisted phones and keeps its membership below a hundred investors. "Few and big is what we have to have," said Thorp. "People come banging on our door, and if they have large enough bankrolls, we take them in." He is vague in public about its assets and actual gambling strategy, saying only that Princeton Newport has accounts in the "many tens of millions" and a return on investment of 20 percent a year.

It was during his final year of teaching at UC Santa Cruz, before he headed south to Orange County, that Alan and Norman discovered that Lewis himself, back in his undergraduate days at Cal Tech, had tried to predict roulette by computer. In 1972 he had loaded a camera case with silicon chips mounted on a printed circuit board. Built before microprocessors were generally available, Lewis's computer required numerous chips to perform its various functions, and it operated at a lower level of sophistication than the machines later built by Eudaemonic Enterprises, but it probably ranks as the first digital computer played in a casino against roulette. Powered by lead acid batteries and operated by two buttons sticking out from the bottom of the camera case, the device gave its predictions by means of a diode display that flashed through a window cut into the top of the case. Like the Eudaemonic system, Lewis's employed a radio link between data taker and bettor. One player manipulated the buttons and then whispered the computer's prediction into an FM transmitter built into his tie clasp. The second player picked up the prediction through an earphone connected to a cigarette pack that was actually a radio receiver, and placed the bets.

Lewis found the best place in Las Vegas to play his computer was Circus Circus. From a second-floor balcony under the high-wire act, he could look down onto the roulette wheels, enter data, and broadcast signals to a partner standing below him at the tables. The program for their computer came from trial-and-error approximations fit on a case-by-case basis. From his balcony overlooking the gambling floor, Lewis in advance of a playing session would clock a roulette wheel with his computer, accumulate data on a miniature paper-tape print-out, and then return to his hotel room to fit a curve to the wheel. After programming the curve into the computer's memory, he and his partner would head back to the casino to play roulette.

"We did some betting," he said. "We didn't make any money, but everything functioned." Lewis's problem, he realized later, lay at the level of theory. At the time he had no idea about the variability of roulette wheels, particularly the differences in their degree of tilt. "Our system wasn't really tested on anything sensible. The existence of tilted roulette wheels is crucial to having a predictive system that works, and we had no idea of it at all. We made other errors and suffered generally from lack of time."

On hearing Alan's story, Norman told him he was working on a more sophisticated version of the same idea, and he invited him over to take a look. "I was surprised to find a miniature commune of physicists," said Lewis of his first visit to the Riverside house. "They were a really interesting and impressive group of people. I liked them all. Just to see a bunch of crazies working on something like this was pretty amazing. I had absolutely no reason to suspect that anyone in Santa Cruz was doing this sort of thing.

"Even the fact that they had such a good roulette wheel was impressive. I knew how expensive they were, and I could tell from that how seriously they were taking the Project. The wheel was a work of art, although by then it had a story behind every burn and scratch mark on it. Given all the time we spent staring at it, the wheel would become for us a great mandala inviting everyone to bow down and worship it."

On joining the Project, Lewis took on the assignment of figuring out a Eudaemonic betting strategy. How big a bank did they require? What percentage should they bet on each play of the game? Should they place their bets on one or more than one number at a time? And what conditions — in terms of rotor speed and tilt — were tolerable for playing a winning roulette computer? To answer the last of these questions, Alan and Ingrid spent several hours a day hunched over the "great mandala" set up on the picnic table in Doyne's bedroom.

"We ran the computer through thousands of trials and compiled success histograms every afternoon," said Ingrid, "before Alan went to the beach. Of all the people I worked with on the Project, I liked being around Alan the most. He had a sense of humor about it and was more relaxed than anyone else. He never viewed the Project as a place to do his stuff and be a star, which is how a lot of us saw it, including me."

On finishing their research, the Lewis-Hoermann team presented their findings at one of the ad hoc Project meetings called whenever anyone had anything interesting to say. "Thorp calculated an advantage for a computer playing roulette of over forty percent," said Lewis. "But the odds are always changing. Sometimes they're higher than that. Sometimes you're actually losing money. And a lot of it's out of your control. But there is in fact a set of ideal conditions for playing a roulette computer. You want a croupier who spins the inside wheel slowly and the ball very quickly, so you have a lot of time to make measurements. You want the table relatively quiet, not hectic, so you have room to place your bets. And you want these conditions to persist for a period of time. This is not an unusual set of requirements, but you're dealing with an uncertain environment. When the conditions are ideal, the odds are quite high in your favor. Who knows what they are? Twenty percent or forty percent or ten percent? You may not know the exact figure, but what you do know is that any advantage of this order, if it persists, will make you a fortune."

No matter how powerful a gambling system may be, there still remains the question of how best to employ it. "If you have a system that really makes forty percent, you still have to ask yourself, 'How should I play it?' There's a lot of literature on optimal systems for favorable games, and the answer is not immediately obvious. Even if you have an advantage, it's clear you don't bet all your money at once. Because what if you lose? You're wiped out. But on the other hand, if you bet too conservatively you may never make any money. So somewhere in between is the right way to play a winning system."

As any gambler knows in his bones, if not in theory, you can have a crushing advantage over the house and *still* get cleaned out of a game. The theoretical analysis of such disasters is known as the probability of ruin. In any contest against Fortuna, favorable odds exist only in the long run, while in the short run statistical fluctuations can appear that look like vengeance incarnate. As Tom Ingerson wrote in one of his epistles to the Eudaemons, "If you're placing $100 a shot bets while looking for a 35 to 1 payoff once in 25 rolls, the statistical fluctuations in the payoffs and the amount of money sunk into investments can get scary. You could easily get into the hole to the tune of $10,000 or $20,000, before seeing any sign that the odds, in the long run, are in your favor."

A chart devised by Allan Wilson, the previously mentioned specialist in biased wheels, and printed in his book *The Casino Gambler's Guide* nicely summarizes the probability of ruin for various games played at different advantages. Posted at the entrance to all Las Vegas casinos, Wilson's diagram might give pause to would-be gamblers. Labeled "Chances of Success and of Ruin in Attempting to Double Your Bank on Flat-Bet Play in an Even-Payoff Game," the chart offers a simple way to calculate how much money you

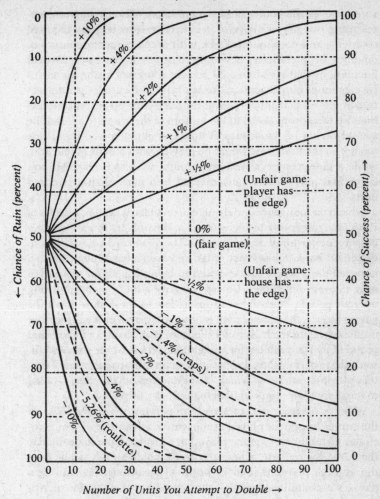

Figure 1. A gambler's probability of ruin.

need to start with, and how you should deploy it, if you want to win a game with a predetermined certainty of success.

To see how the chart works, take as an example the average Thorpian card counter who plays blackjack with a 1 percent advantage over the house. Find the curve on Wilson's diagram representing this player's edge of +1 percent. Now let's say the card counter wants to play blackjack with an 80 percent certainty of doubling his money. (He is willing to expose himself to statistical fluctuations that give him a 20 percent chance of losing his shirt.) He calculates his proper bank-to-bet ratio by locating the intersection of the +1 percent curve and the horizontal 80 percent success line. He then drops down to the bottom of the graph to see that he should divide his bank into 70 units. A gambler betting all his money on the first draw would be playing with a bank of one unit, while a more conservative player might wish to extend Wilson's curve and play with a bank divided into more than a hundred units.

When you look more closely at Wilson's chart, certain aspects of it become unnerving to the professional gambler. To gain only an 80 percent certainty of success, the Thorpian card counter has to divide his bank into seventy units, and even a bank divided into a hundred units leaves him exposed to being wiped out more than once in every ten games. Unless you have a lot of money burning up your pocket, bank-to-bet ratios as low as these slow down the game, reduce the chances for a "big win," and prove altogether uninteresting to the high roller ignorant of probability theory, and proud of it. The problem for roulette players is worse still. As Wilson's chart pertains to "flat-bet play in an even-payoff game," bettors playing roulette — which is not an even-payoff game — need to multiply their required banking units by thirty-five.

Wilson's diagram becomes truly unnerving when one looks at the curves for games played at a negative advantage — which includes virtually every game played in a casino. Note in particular the curve for roulette, where the house holds an advantage over the average player that looks positively usurious. On reading a player's probability of ruin — this time from the left side of Wilson's chart — a startling fact comes to light. The more units into which the average roulette player divides his bank, and the longer he tries to stay in the game, the *greater* his probability of ruin. This leads to the following conclusion: "If you want to double your bank on an American roulette wheel," says Wilson, "there is *one and only one best bet*. You must take *all* the money that you intend to risk at roulette during your whole life and shove it all down on one spin of the wheel on one of the even-money propositions."

A computerized roulette player with an advantage of 40 percent can approach the game with more creative leeway, but even an edge as large as this entails a slight possibility of having the odds run against you long enough to wipe you out.

"Under the best of all possible conditions," said Doyne, "the computer's advantage in roulette is something like a hundred percent. You can predict where the ball will come off the track three revolutions ahead and almost always pick the right half of the wheel on which it's going to land. A hundred percent advantage means you have a two-to-one edge over the house. If your prediction allows you to eliminate half the numbers on the wheel, the number you're betting on is twice as likely to appear as ordinary, and you thereby double your money.

"Because of bounce and scatter and other conditions that are less than optimal, the computer's real advantage is more like forty percent, which means that we're shadowing, or eliminating, eight unlikely numbers from the wheel. In normal play this is a fair guess of what actually happens, and it agrees with what Thorp and Shannon found."

The greatest possible advantage in roulette, which comes from predicting with absolute certainty one of its thirty-six numbers at each play of the game, equals thirty-six times one hundred, or 3600 percent. "Only the ultimate roulette scheme would give you a thirty-six-hundred percent advantage," said Doyne. "That's where you're using a psychic, who can peer into the future and then roll time backwards."

Intending to spend the summer in Las Vegas playing roulette, the Projectors devoted the spring of 1978 to debugging equipment and testing the system. They planned to work in rotating shifts. Having learned a lesson from the debacle of their recent trip to Reno, they would keep these groups small, well equipped, and highly trained. "We had our eyes fixed on the Nevada lettuce patch," said Norman, "but we were going to be very crafty about getting in there to eat out of it."

"Once the equipment was working reliably," said Doyne, "we wanted to test the boundaries of the program so that we knew exactly how far we could push the computer. We were going to explore the envelope. The idea was to go to Las Vegas with a number of trained data takers, increase our stakes, and play the thing out."

Scheduled to leave Santa Cruz as soon as the term ended, the Projectors were delayed nearly a month while debugging equipment. Dave Miller, a new recruit to roulette madness, came from Silver City to help out with what everyone had taken to calling "the hardware wars." A college engineering major and former member of Explorer Post 114, Miller had gone on to achieve local fame as New Mexico's motocross champ.

Having last seen them in Las Vegas, Tom Ingerson also arrived during the hardware wars to find his former Explorers possessed by roulette. Even while jogging along the San Lorenzo River, all they could talk about was computers and gambling. "He wasn't what I expected," Ingrid said of her first meeting with Ingerson. "Instead of a young, dynamic person — a real initiator in getting people to do things — he seemed like anyone else in his forties. Lonely and a little bit insecure, he was starting to develop the habits of older people. Things had to be a certain way. He got nervous around messes. He was put off by the house feeling like a camp or a football team, with everyone getting psyched up to go out there and meet the enemy. We were all using the same jargon and had the same glint to our eyes. He couldn't get us to talk like normal people and show respect."

Ingerson helped Norman fine-tune the receivers, while other visitors to the Riverside house got drafted into Eudaemonic service debugging and testing computers. "By the end of the hardware wars we had worked ourselves up into a pretrip frenzy," said Norman. "This was mingled with an exhaustion that rotted judgment."

In the middle of June, Alan, Ingrid, Doyne, and Norman, composing the "first wave" of summer roulette players, loaded three vehicles with five computers and set out from Santa Cruz to cross the mountains and desert to Las Vegas. For transportation, they had Alan's Triumph, along with the Blue Bus and Dave Miller's VW van, both of which were packed to the ceiling with roulette gear and all the pots, pans, and other utensils needed for spending the summer in Las Vegas. For computers, they had Harry, Patrick, Renata, and Cynthia, as well as the KIM and its PROM burner, in case any last-minute changes were required in the program.

Heading due east from Monterey Bay into the Sierra Nevada,

they spent the first night camped in the Motherlode along the Merced River. "Then we marched across the Sierras in a little caravan," said Norman. "It was my first trip to Yosemite, and we spent a beautiful second day driving up to Tuolumne Meadows. Green and filled with mountain streams, it felt like a fairyland on top of the world."

From there they made their way over Tioga Pass and began the long descent into the desert. Leaving behind the pine trees and meadows, they dropped down the arid flanks of the eastern slope toward Mono Lake, whose reflection glimmered on the horizon like a vast inland sea. With the sun burning overhead, they were looking forward to stopping for a swim. But the blue mirage kept receding in front of them, and when they finally reached the shore, they found only the pathetic remnant of a lake long ago sucked dry by Los Angeles. "It was nothing more than a mud flat with a little puddle in the middle full of brine shrimp," said Norman. "But we were hot and tired, so we parked on the edge of the mud flat and walked out to the puddle to go swimming."

Afterwards Ingrid felt sick. In the past few days she had been falling asleep at odd times and forgetting things. The Bus was already halfway to the Nevada border when she remembered leaving her wallet at Mono Lake. They turned back, only to discover later that the wallet had been in Ingrid's backpack all along. "I got angry," said Doyne. "I thought her being sick was psychological. I was feeling the pressure again, thinking to myself, 'Are we ever going to get over the hump and do it?' I was slave-driving everybody as hard as myself."

Out in the desert on the outskirts of Tonopah, Dave Miller's van threw a rod. It was night, and no stores were open to buy even a chain or a tow bar. At a Chinese restaurant in Tonopah, Doyne found some cowboys wearing snap-button shirts and ten-gallon hats. They didn't have any proper rope, but one of the cowboys sold him a lariat. It was really too short for towing, except in a pinch. Norman voted to spend the night in Tonopah. Alan Lewis was noncommital. Ingrid was asleep. Doyne insisted they shove on. After tying the van to the Blue Bus with the lariat, he and Norman worked out a system of signals for slowing down and stopping. Maybe Doyne saw the signals and ignored them, or maybe the lights of the van were too close to be seen from the Bus, but he maintained a steady sixty-five miles per hour towing Norman at the end of a five-foot lariat against his will through Scotty's Junction, Lathrop Wells, Indian Springs, and on south through the Sonoran Desert to Las Vegas. When they stopped on the Strip, Norman's face was white with anger.

"Usually when you make decisions with Doyne he appears to be completely rational. But this time he was just imposing his will."

Arriving in the early hours of a June morning, they found the temperature in Las Vegas within striking distance of a hundred. They drove to Len and Jeri Zane's house — on loan for two weeks while the Zanes were out of town — and unloaded their gear.

"In no time," said Norman, "we had the living room and the rest of the house transformed into a junkpile of electronic parts." When not working on the reflex tester or tweaking up equipment, the Projectors studied casinos for likely wheels. They compiled tilt histograms. They sketched floor plans and inserted them in the black three-ring notebook with separate entries for each casino. They mapped the location of bathrooms, exits, cameras. They recorded shift changes and made notes on the style of individual croupiers.

For sauntering around town, Norman bought a polyester leisure suit. "We had a problem," he said, "in needing sports coats and sweaters to hide the computer, but here we were in the desert in the middle of summer, and it was a hundred degrees in the shade. We solved the problem by affecting the classic Las Vegas look: the leisure suit, consisting of a lightweight polyester jacket with matching pants in pastel colors. Alan Lewis already had a polyester wardrobe, but the rest of us had to scavenge through Goodwill and the secondhand shops." Doyne sported a Mexican wedding shirt and a pair of purple knit pants inherited from Norman. Ingrid once again did her best to look alluring in her floral-print rayon blouse worn with ratted hair, makeup, and computer-padded bra.

Like other professional gamblers, the Projectors breakfasted after noon and arrived at "work" by the cocktail hour. A typical day might begin with a ninety-nine-cent breakfast special at the Golden Gate, followed by a stroll through the casinos on Fremont Street to eyeball wheels and jot notes on those that looked promising. Because it simplified the process of roulette prediction, choosing the best from among the town's many nicely tilted wheels was worth the effort.

The Projectors scheduled two gambling sessions a day, one in the late afternoon and the other from midnight to three in the morning. This was a particularly good time to play roulette, as the casinos were relatively quiet, while a lot of wheels that would stop with the graveyard shift were still in play. Shift change times varied from casino to casino, and information such as this was carefully recorded in the black notebook. "Ideally," said Norman, "we didn't want to play too long on one shift, to avoid drawing attention to ourselves. So we played an hour or two on one shift and then took a break before playing the next."

The two-person system was now perfected. Wired with the data-taking computer and transmitter, Doyne, Alan, or Norman would approach a wheel and start clocking the ball and rotor with toe clicks. Stopping along the way to adjust the program's variable parameters, they required roughly twenty minutes to drive the computer around the mode map. They then clicked into the "play" mode.

Having given the data taker a head start, Ingrid or another bettor would arrive at the table wearing the second computer and receiver. The two players were connected primarily by radio link, but they also "talked" to each other by means of signals on the layout. A chip placed on red or black, odd or even, or the various side bets in the game might mean "take a five-minute walk," "sit down and play," or "raise stakes."

It was now second nature for the four of them to walk around town wired from head to toe with microswitches, antennas, battery packs, and computers. While pumping their toes and counting solenoid buzzes on their stomachs, they could still manage to crack jokes and flirt with croupiers and hostesses. Doyne in particular was a master of casino disguises, adept at keeping his face as wide open and ingenuous as the prairie out of which he must have come. Under normal conditions, this face was animated with thought, but here in the casinos it was cleansed of expression, and those parts of it that were slightly akilter — a chipped tooth, a nose bent at the base from having been broken three times, a lopsided smile — were used to construct an impenetrable mask. Doyne resumed his southwestern twang and slipped back into the persona of New Mexico Clem, the poker sharp last seen at the Oxford Card Room in Missoula, Montana.

The Eudaemonic system demanded role playing from opposite ends of the spectrum. The data taker, while standing next to the wheel, made only insignificant bets and developed other techniques for self-effacement. The bettor, standing farther down the layout, never looked at the wheel and placed high-stakes wagers straight up on individual numbers. Between these simple differences lay an immense psychological gulf. "There is a sharp distinction here," wrote Dostoyevsky in *The Gambler*, "between the kind of play that is called *mauvais genre* and the kind a respectable person can allow himself. There are two kinds of gambling, one that is gentlemanly and another that is vulgar and mercenary, the gambling of the disreputable. . . . A gentleman, for example, may stake five or ten louis d'or, rarely more; he may however stake as much as a thousand francs if he is very rich, but only for the sake of gambling itself, for nothing more than amusement, strictly in order to watch the process of winning or losing; he must by no means be interested in the winnings themselves. When he wins he may, for example, laugh aloud, or pass a remark to one of those standing nearest to him, and he may play again, and then double his stake once more, but solely out of curiosity, to observe and calculate chances, not out of any plebeian desire to win."

According to Dostoyevsky's model, the Eudaemonic data taker seemed a disreputable, if harmless, player of *mauvais genre*, while the Eudaemonic bettor displayed only the most "gentlemanly" of attributes, including an aristocratic disdain for the spinning wheel and workings of the game. The data taker actually operated under an altogether different and, in fact, paradoxical set of constraints. His plodding attentiveness to the wheel had to be *exaggerated*, so that the drudgery of it appeared as something stupid or harmless.

Three kinds of people in Las Vegas stare at roulette wheels: (1) rubes off the Great Plains, (2) people who have never before played roulette, and (3) system players. Doyne, in his virtuosity, could look like any or all of the above. He could stand next to a wheel for two hours with a computer tucked under his armpit and be completely ignored by croupiers, wheel spinners, bankers, pit bosses, floormen, and cocktail waitresses — people trained expressly not to ignore *anyone*. Looking like the Son of Sam tuned in to voices from other planets, Doyne was an obvious case of someone with a screw loose. At other times he affected the look of a system player. He would take a pencil from behind his ear and doodle numbers in a notebook, thereby joining the well-respected company of those ignorant of the fact that no mathematical system for beating roulette exists.

Only once during the summer did someone suspect a possible connection between the Project's data taker and bettor. "Doyne and I were playing together," said Ingrid. "He ordered an orange juice from the cocktail waitress, and I asked for a ginger ale. On the next round we reversed orders. I asked for the orange juice and he ordered the ginger ale. 'Are you two together?' said the cocktail waitress. 'No,' I explained. 'It just looked real good.'"

There was another occasion when someone remarked on the Project's style of betting. The computer output its predictions via octants containing four or five numbers clustered together on the wheel. These proximate numbers on the rotor, when rearranged on the layout, lie scattered across the length of it, and only a sophisticated player would note the relationship. To further disguise their betting patterns, the Projectors varied the makeup of octants by substituting adjacent numbers.

Despite all precautions, there was one winning session at the Holiday Inn when two bystanders remarked that Ingrid was betting by octant. "They were college students in town to play blackjack, and they had 'card counter' written all over their faces. They watched me play and then asked if I had a system. I made up a story about progressions, but then one of them figured out I was betting by octants and said so out loud. Of course I had to cash out and leave the game."

The Projectors traded among themselves the roles of data taker and bettor, but neither Alan nor Norman could efface himself as idiotically as Doyne. Norman, pushing over six feet, bearded and lean as a coconut palm, could never completely drain the intelligence out of his face. Looking natty in his leisure suit and pastel pants, his most workable persona was that of a system player. Alan Lewis, dark-haired and tanned, seemed completely at home in Las Vegas. His parents had lived there for a short time, and as a former card counter, he knew his way around the casinos. With his Triumph Spitfire and wardrobe of stay-pressed slacks and drip-dry shirts, he needed only a few gold chains around his neck to look the classic Las Vegas high roller.

Over the course of the summer of 1978 the Eudaemonic teams staked out favorite casinos and tables. In the sawdust joints on Fremont Street the wheels were nicely tilted, but the croupiers out on the Strip, especially late at night, played the kind of leisurely game picked off most easily by a computer. Like commando teams with disguises, prearranged signals, and sophisticated equipment, the Projectors circulated night after night from the California Club to the Lady Luck, the Golden Nugget, and El Cortez before heading out to the Strip for a raid on Circus Circus, the Silverbird, and the Stardust, or sessions farther to the south at Caesars Palace and the MGM Grand.

They never played the same club on consecutive nights. They entered casinos separately and "talked" to each other only via signals on the layout. Each of them had been issued a small notebook for recording data and financial transactions. At the end of the day this record of wins and losses was transferred to the black notebook, which had pages listing each casino separately, and a master page devoted to keeping track of the Eudaemonic bank. Entitled "Daily Record," the master page was ruled with separate columns for recording the date, casino, shift, data taker, bettor, average bet, number of trials, money won or lost, time spent at the table, and running Eudaemonic balance.

"It was like being in a war," said Ingrid, "and the characteristics that were important in people were the kind that seem important in wartime. You had someone like Doyne who tried to push ahead, no matter what the obstacles. And then you had a person like Norman, the dependable guy who stayed up all night putting in the hours that it took to get the equipment working properly. And then there was me, who tried to make everybody feel comfortable. That was my role — sort of a mascot."

Drowsy and forgetful ever since her afternoon swim in Mono Lake, Ingrid was examined by a doctor and declared to be suffering from mononucleosis. "I went upstairs and went to sleep and never woke up. My dreams were filled with water, and my memory of the whole summer is of being held underwater. I had worked myself into a frenzy over the Project, trying to be supportive and keep the

energy high, until I finally gave out." Ingrid woke up long enough
to read Dashiell Hammett novels and play occasional games of
roulette. She spent July recuperating at her parents' house in
Davis, and returned to Las Vegas in August to play a heroic role in
the summer's final assault on the casinos.

Like the city's other professionals, the Eudaemons on their
nightly rounds confronted moments that later took on legendary
significance. One of their favorite casinos was Circus Circus. With
tour buses surrounding it ten deep, a flying trapeze over the main
floor, and slot machines on a merry-go-round, this Strip casino caters to families looking for wholesome gambling. It also has tilted
roulette wheels that tend to be played at a leisurely pace.

Alan Lewis knew the casino well from having shadowboxed his
own system there in the early 1970s. Returning now with a
smarter and more sophisticated computer tucked under his arm,
he was second only to Clem from New Mexico in his skill as a data
taker. "Doyne," said Alan, "had developed an amazing ability to
talk to croupiers and simultaneously keep track of where he was
in the program. You had to time accurately and be comfortable
with what was going on. You had to concentrate and not get sloppy
about things. Working the microswitches with your toes was not
so easy that you could casually do it, in which case you'd be so
inaccurate as to be useless. That's why the reflex tester was an important element in planning the whole thing. It was a nice touch
to the Project."

It was at Circus Circus that everything clicked into place for
Alan's first big win. Trying to look like a system player willing to
count numbers until the sun came up, he stood next to the wheel
as data taker, while Norman, rumpled but perky in his leisure suit,
played bettor. They had been losing money for some time. Alan
was having trouble setting parameters for the wheel, and he was
about to call off the session when a shift change at two in the morning brought on a new croupier, a woman who already looked bored
and ready to turn in for the night. In the early morning quiet, as
she spun the rotor and launched the ball up on the track, the game
in her hands took on a wonderful precision and regularity. Alan
fine-tuned the parameters and got the computer clicked into playing the game.

"Things just turned around," he said. "The predictions were

right on target. Playing quarters, we recouped our losses and stacked several hundred dollars in chips in front of us. There was a gut rush of excitement in seeing everything fall into place. After all the time spent testing and troubleshooting it, the computer was finally up and running perfectly. I no longer had the slightest doubt. From that session, I knew the game of roulette had been beaten."

9

Lady Luck

Do not approach casinos with timidity or reverence.
They are simply fruit-machines tended by bank clerks
and mechanics.

Ian Fleming

With the Project's computers performing exactly as planned, the casinos looked as if they were getting worried. A croupier at the MGM Grand — mystified and upset by the stack of chips piling up in front of Ingrid — broke a fundamental house rule by grabbing the wheel and wrenching it around on its seat while the game was in play. The wheel made a tremendous screech. The shift boss came over to ask what was going on. In no position to lodge a complaint, Ingrid cashed out of the game and left the casino.

At another winning session, a croupier interested in Norman's system asked him what she could do to help. Avoiding the technical term, Norman said, "Do you mind spinning the ball a little harder around the outside of the wheel? That makes me feel luckier." Yet another croupier offered the Eudaemons an elegant demonstration of what gamblers call the dealer's "signature." Imagine someone operating a roulette wheel five nights a week ten years in a row. The slight imperfections on the track, the characteristics of different kinds of balls, the heft of the rotor and its drag on the central spindle will become intimate facts of life. On a slow night in the casino, to stave off boredom, what if our imaginery croupier experimented with the game, looking for ways to enhance the regularity and precision of it? Like a relief pitcher perfecting a slow sinking curve, the croupier after years of practice might learn how to flick the ball up on the track and drop it from orbit precisely twenty revolutions later. What if he then learned how to regulate the speed and position of the rotor, until, with several more years of practice, he perfected a synchronous loop in which the ball arced neatly on its twentieth revolution into a chosen pocket waiting below?

Doyne and Norman were teamed up to play roulette at the Lady Luck, a favorite casino of theirs because of the free tuna salad sandwiches and two-egg breakfasts served twenty-four hours a day. The croupier that night, a man in his thirties with frizzy hair, was particularly friendly. The computer was also working exceptionally well, so that early in the game Norman had accumulated several hundred dollars in chips. Standard protocol in roulette calls for winning players to tip the house, and Norman was doing so by giving the croupier chips to bet at his discretion. The man always chose number seventeen, which was odd, considering that the computer itself consistently predicted the octant of numbers containing seventeen.

"Why do you always bet on number seventeen?" Norman asked.

"Because if I do everything just right," said the croupier, "I can actually hit seventeen. Not that I can do it every time, mind you, but I can get close to it. I set the wheel going at a steady rate, and then I flip the ball in this nice regular way when the zero is lined up in front of me, and I swear I can hit seventeen with better-than-average odds."

On his last spin of the wheel before going off duty, the croupier placed his tip on number seventeen. The computer predicted the octant holding seventeen. Even Doyne as data taker got excited enough to throw a bet down on seventeen. Twenty revolutions later the ball landed right on target. "That night," said Doyne, "we all went away happy."

The summer's most frightening experience came during a winning session played by Doyne and Ingrid at the Hilton. For bystanders unaware of its existence, there was something uncanny about the computer at work. Placing bets straight up on three or four numbers at a time, Ingrid was "guessing" the winning number with surprising frequency. Stacks of chips mounted in front of her, and she tried to look surprised each time the croupier pushed another pile in her direction.

"Some days you have it, and some days you don't," she said to the other players at the table. "I guess this is my lucky day."

Suddenly there appeared on either side of her two large men in suits. They stared down at her. One of them wrote in a notebook. They pressed against her and waited for her next move. Ingrid placed a chip on 00, the signal to quit, and cashed out of the game. "There was so much paranoia involved in playing the system," she said, "that I could be making all this up. But when I later went into the coffee shop, I saw the same two men sitting together at a table, and I'm sure they were house cops.

"We had agreed that anyone feeling heat could call off a session, even if the other person hadn't felt it. We realized we were still weak on basic things like disguise, for which we hadn't come up with a good solution. Even though we pretended not to know each other, here were two people roughly the same age standing at the same table. I was afraid it wouldn't take them long to figure out something was going on. And as soon as they did, I realized how easy it would be for them to do something awful to us."

On surfacing from her underwater dreams and walking into the equally dreamy landscape of Las Vegas casinos, Ingrid imagined making a movie of the Project. "I thought of a movie playing out people's fantasies on the screen. In a collage of images focusing on our perceptions and paranoia, I wanted to get at the weirdness of Las Vegas from the inside.

"The Project for me was like a theater or music performance. It was like the 'happenings' I had been doing at school. But the Project was a performance in real life and for a reason: to make money underground without having to work a normal job. There was an element of danger, but it seemed like just the right amount. What you were paranoid about was the real unknown."

Doyne himself began making notes in a "Movie Idea Book," one of whose early entries reads, "We're either going to end up with a hell of a good movie or a hole in our heads." One imaginary scene depicts a fat Argentinian smoking a cigar and losing two thousand dollars in front of an admiring crowd. A woman tries to steal chips from Ingrid while bragging about having lost her ex-millionaire former husband's fortune. The Argentinian tells the woman to be happy, that her luck is bound to change. The croupier flashes his diamond pinky ring for everyone to admire, and then in the next frame Doyne falls writhing to the floor after having been electrocuted by a short circuit in the computer. He rips off his antenna T-shirt and passes out in the arms of the pit boss.

Under the heading "Themes," the Movie Idea Book contained notes such as the following: "Must have lots of us — minor characters, tourists — putting in their two bits about everything. Juxtapose these monologues to illustrate basic themes: adventure story, surrealism of gambling, money and capitalism, dreams and fantastic schemes. Have Hunter S. Thompson babbling on ether appear in Las Vegas with Samoan attorney and white Cadillac. Study psychology of gamblers. Contrast with our own motives (why we crave adventure, fame, money) to show how in many ways we are motivated like the gamblers who throw away their money."

Scripted to the music of Pink Floyd singing "Money," the movie would describe in its final scene how "We fix our computers, get the Project going, move to a desert island, and build rockets to travel to another planet. . . ."

When the Zanes returned from their vacation in July, the Projectors moved into a two-bedroom apartment on Tropicana Avenue — the main road running east from Las Vegas to Lake Mead. "Norman and I gave the manager a screwball story about our being consulting electronics engineers," Doyne said. "He wanted respectable tenants, so we couldn't tell him we were in town to play roulette." The landlord in turn neglected to mention that living in the apartment above them was a speed freak who paced the floors in high-heeled shoes while pimping to a select clientele of truck drivers, cops, croupiers, and anyone else who knocked on his door.

After moving into the apartment, the Projectors played roulette for another week and then stopped to take a break. Lorna Lyons and Rob Shaw rolled into town in his "Cream Dream," a white 1959 Ford station wagon. They had come to pick up Norman and drive him back to Santa Cruz. Ingrid went home to Davis to recuperate. Alan Lewis flew to see his family in Tucson. Doyne headed for Letty's house in Santa Monica, where she was living with friends near the beach.

He arrived with all the computers, the KIM, the PROM burner, and enough spare chips to reburn the computer program, because he had thought of a number of bright ideas for modifying the algorithm. He imagined he could simplify the process of setting parameters and thereby shave a few minutes off the time needed to drive around the mode map. Doyne programmed in the morning and body-surfed off Venice beach in the afternoon, and two weeks later, on reburning the EPROM — the erasable programmable read-only memory in which the microcomputer stored its equations — he was able to run the computer around the mode map in record-breaking time.

A new group of Projectors gathered in Las Vegas at the end of July. Arriving from Santa Cruz via Los Angeles, where they had stopped to pick up Doyne, were John "Juano" Boyd, Marianne Walpert, and Chris Shaw, Rob's brother. Chris was an artist working with Ralph Abraham on a series of "visual mathematics" books, including one on chaos and strange attractors. But he was also a *bon vivant* whose suavity Doyne thought would lend itself nicely to high-stakes betting. Driving out of Los Angeles at the wheel of her blue Comet, the irrepressible Marianne rear-ended another car on the Santa Monica freeway. When Chris jumped out, clambered up on the hood, and popped it back into place, it dawned on Doyne, "that I was heading for the Strip with three untrained players, and frankly I was skeptical.

"I was especially worried about Juano, who had hair down his back and a neck beard. He was wearing the same brown plastic glasses he'd worn his freshman year in high school, except that now they had a lot of tape holding them together. I didn't see any way he could make it in the casinos."

Juano submitted to a haircut from Chris and took other hints on how to straighten up his image. The three untrained players set up the wheel and reflex tester in the Tropicana Avenue apartment, while Doyne worried over sporadic glitches in the receivers and last-minute programming changes. It was summer in the desert. Too hot during the day even to go swimming, because you burned your feet on the concrete around the pool.

"I was struck by the insanity of it," said Marianne. "We were living in an apartment full of computers and electronic gear. We were camped in Las Vegas in the middle of summer, and sometimes I got angry at myself. I could have been doing something fun, and here I was hanging out in this incredible heat or in stupid, goddamned casinos. The rest of the time I was sewing snaps onto brassieres or wiring myself from head to foot. You really had to wonder sometimes why you were doing it. But the amount I questioned it must have been nothing compared to Doyne and Norman, who had already given up years of their lives to work on the Project."

The Eudaemons cooled off in the evenings at Roulette Rapids, an amusement park with streams of water flowing down cement chutes. Sitting on foam rubber mats that scooted down the tracks at high velocity, people tried not to fall off for fear of getting bruised on the concrete. When not tinkering with hardware and software, Doyne read Joan Didion's *Play It as It Lays* and Richard Dawkins's *The Selfish Gene*. The hit song that summer, played over and over again on the radio, was something called "Hot Blooded":

> Well I'm hot blooded
> Check it and see
> I got a fever of a
> Hundred and three

At the end of the first week Letty flew in from Los Angeles. She learned the betting pattern in a couple of hours, and then she and Doyne went to play roulette at Circus Circus. Initially, she was skeptical about her ability to take part in the Project. "If I go into a casino as myself," she said, "I don't fit in very well. I feel cloddy and out of it. And if I go as anybody other than myself, then I have to act, and I'm not an actor. I've always liked the idea of acting and admired actors, but it's never been anything I had any talent for." Her only prior experience on stage had been as a WAC in her ninth-grade production of *The Mouse That Roared*.

On her maiden appearance as a Eudaemonic bettor, Letty bought into a game and stacked her chips, before realizing that she had forgotten to turn on the computer. For men the power switch was in their pockets, but for women it was under their shirts. She excused herself for a visit to the toilet and returned to play a winning game of roulette. She proved a natural at using the computer. She was cool under fire, precise and attentive. The next night at the Holiday Inn she logged another winning session.

Letty had been what she described as "pillowcase adviser to the Project," but now she also made some trips to the library to research the legality of beating roulette by computer. Although it is against the law in Nevada to tamper with the outcome of a game, *predicting* its outcome — by whatever means — is perfectly legal. Recent court cases have confirmed this fact in the instance of card counters. Everyone knows, though, that no matter how legitimate it might be to walk into a casino wearing a roulette computer in your bra, the bosses will undoubtedly think, and act, otherwise.

Apart from the danger of it, Letty had other doubts about the Project that were moral in nature. These had to do with social justice and right action, and they sprang, as she freely admitted, from her family background. A Boston lawyer of the liberal persuasion, her father balanced charitable work for civic groups against the bread-and-butter lawyering of trusts and estates. "My mother is of the old-fashioned school. She's always maintained that politics is the highest calling." While working in admissions at Radcliffe, Letty's mother also served as fund raiser and board member for numerous social causes.

"I think there is a liberalism," Letty realized, "that can be bred by growing up in a situation where you never need money. Making money as a justification for doing something was never legitimate. You had to explain your actions some other way. At times I yearned not to have any money, because then all I'd have to do is figure out how to make it, and life would be easy.

"When you aren't lacking for them, the purpose of life is not to acquire things. Then what is the purpose of life? You fool yourself into thinking that it must be to help everyone else along. I was bothered from the beginning that the immediate purpose of the Project was to make money. Maybe it's just my do-good ethic, but I always wanted a greater goal. Not that I ever thought the Project was immoral. It's just that it didn't promote some greater sense of morality, my own morality at least."

In spite of her qualms about it, Letty understood the importance of the Project to Doyne. "Doyne has always felt that he's lived a century or two or ten or a hundred too late. He's yearned to be an explorer, an adventurer, an individual fighting adversity to build something he cared for. For Doyne the goal of the Project is money, which is freedom. That's the immediate goal. Then he can go do things he wants to do. In the end he wants to be free in spite of society, the government, corporations, respectable people, and all the other upholders of order who say, 'You can't do this sort of thing. We want you on this road over here. You're a nice respectable type with a Stanford physics degree, so why don't you just join us?'

"The Project seemed a way for Doyne to make it on his own. There's a big pot of gold at the end, and you, only you, have figured out how to get it. That you could beat these huge forces in a world designed to get every penny and not let a single one out, that you, just one person and your friends and associates could beat that system, that's very appealing. And to do it from scratch, with no money, without selling your time or giving people promises, to do it through your own sheer wit and determination, to put yourself out there instead of being a passive cog in someone else's wheel — that's the attraction of the Project.

"Of course it has larger goals beyond making money. These involve working with friends to accomplish this very difficult task. It's wonderful to think of rounding up everyone you know and using all their ingenuity and scientific expertise to work on a secret scheme, which you then have the fun and adventure of pulling off. It's like doing a play together, where everyone has a role, either onstage or behind the scenes."

After Letty's winning sessions at Circus Circus and the Holiday Inn, the second wave of the Project rolled back from Las Vegas for another pause. They needed a break after a week of speed freak pimps, desert heat, and roulette computers insincere about whether they were on strike or just being flaky. Marianne and Chris headed for the Coast. Letty flew back to Los Angeles. Doyne and Juano were picked up by Tom Ingerson, who drove them south to Kingman, Arizona, where Juano was dropped off to hitchhike to Mexico. He got as far as Barstow before being robbed, stripped bare of everything including his eyeglasses, and deposited in the middle of the desert. It was a tale he lived to tell, and the culprits were later caught by the police, but Doyne's sense of unease about Juano had proved prescient.

A reunion had been organized for the old Explorer Post 114, but in advance of the gathering Doyne and Tom spent three days camping together in the Gila Wilderness. During their conversations about the Project, Tom was stern. He thought it was a sinkhole of time and energy. He believed the technical difficulties would swamp them. He considered the dangers too great to offset any expected return. "He basically told me," said Doyne, "that I was wasting my time."

The third wave of Projectors — with Doyne and Norman driving up from Silver City, Alan Lewis flying in from Tucson, and Ingrid taking the bus from Davis — converged on Las Vegas early in August. They picked up a newspaper and drove around town looking for another apartment to rent. They settled on a two-bedroom walkup in one of the sleazier neighborhoods near the Showboat casino. The apartment was decorated in luau pink. Too many chain smokers had coughed up their lungs in these rooms. Down and out gamblers had probably done worse. The place had an irremediable air of lovelessness and transience, but the price was right. They moved in and within an hour were practicing on the wheel and biofeedback machine.

The allure of Las Vegas lies in money that you can finger. The New York Stock Exchange handles more of it every day, but the money there has been abstracted into certificates or digitized into read-outs. The money in Las Vegas is tangible and fluid. It rises in waves over the tables, eddies into whirlpools, and gets sucked again into the great sea of money that washes back and forth over the casino floor. Because these chips, coins, bills, and silver dollars slip so freely over and under the tables, a lot of people can be found walking around Las Vegas with money stuffed in their pockets. Consequently, as one would expect, there are a lot of other people who specialize in unstuffing these pockets.

Doyne had a chance to meet one of these specialists soon after moving into the Showboat apartment. "I frequently have trouble falling asleep. It was three in the morning and I was just on the edge of unconsciousness one night when I saw a figure in the room. 'That's strange,' I said to myself. 'What's Norman doing wandering around at this hour of the night?' 'Norman?' I called, and suddenly the person, carrying a piece of clothing, bolted out of the room.

"I jumped up and ran after him down the stairs. I was stark naked, and my penis was flopping around as I sprinted down the street at top speed, yelling, 'Stop, thief!' The sidewalk was littered with broken glass, and I normally wouldn't have walked on it barefoot, much less run.

"About six feet tall, the thief was wearing cut-offs, a T-shirt, and track shoes. He was in good shape and obviously knew the neighborhood. But I was in good shape, too, having run five miles a day that summer. Going at an all-out sprint, I was gaining on him, and when I got within fifteen feet I screamed, 'If you don't drop those pants' — because it was my pants he'd lifted from the bedroom — 'I'm going to kill you when I catch you.'

"My mother had given me five hundred dollars for a stake, and my brother had put up another five hundred dollars. That was the bank. So there was a thousand-dollar debt right there in my pants' pocket that I couldn't see any immediate way to pay off. Plus there had been another six or seven hundred dollars in cash in the room, along with traveler's checks, chips, and silver dollars. When I had almost collared the guy, he veered into an apartment complex with locking gates, and I finally lost him where he could have headed in any number of directions."

On returning home, Doyne discovered that one of the neighbors had called the police, who were soon at the door. They walked in to find the living room filled with antenna T-shirts, solder guns, and chips. "Someone had covered the roulette wheel with a sheet, but it was obviously a pretty weird operation. We gave them a story about being college students working on a summer research project in electronics. It turned out that all I had in my pants was two dollars and Ingrid's driver's license, which she had already lost several times that summer. But the incident scared us into keeping things better under wraps."

The Projectors paired off into teams and tried to play both sets of computers at once, although it was hard to keep all four of them in working order. With radio receivers and wires perennnially on the blink, Norman and Doyne functioned more as repairmen than gamblers. They filled the Blue Bus with tools and stationed it in strategic parking lots between casinos. Alan Lewis was having difficulty mastering the revised program, and he lost a few hundred dollars at Circus Circus before learning his way around the new mode map. The Projectors in general found themselves breaking up a lot of sessions because of malfunctions, shocks, spurious buzzes, and other hardware problems.

"Ingrid and I were doing better than the others" said Doyne. "Without any fantastic successes, we were winning at a steady rate." The pattern changed suddenly one night when they were playing roulette at the Lady Luck down on Fremont Street. "We found a slow wheel with good tilt," said Doyne. "I had set the parameters just right, and we were basically ready to kill them. So we dug in and started playing. But we were the only people at the table, and we were getting a lot of heat from the pit boss, more than we had ever had before. And Ingrid was acting strange. She was jerky and nervous, and I couldn't figure out what was going on."

"The pit boss," said Ingrid, "was a tall, weasely fellow who must have sensed we were nervous. I was acting strange only because I was getting shocked by the antenna wire over my left breast. The shocks came more and more often, until I got muscle spasms and had a hard time remembering the betting pattern. But we were winning a lot of money; so I thought I should keep going."

"We made several hundred dollars in our first few minutes at the table," said Doyne. "When you hit successive throws, the money comes extremely fast, and this time it was quick, really quick. Even with perfect parameters, there are statistical fluctuations that keep you from winning every time. But when the conditions are good, the winning predictions can come right on top of each other. Suddenly I saw two plainclothes detectives standing off to my left in front of the wheel. I gave Ingrid the signal to lower the bets. But she *raised* them instead, and we hit on a five-dollar chip and then another. She was pushing the stakes hard.

"Standing next to me were these large men with potbellies, and I overheard one of them say to the other, 'You see that woman betting on number nine? I could swear she knows where the ball is going to come off.' They obviously didn't realize I had anything to do with Ingrid, but I cashed out and left anyway."

When they met later in the parking lot, Doyne discovered why Ingrid had been acting strangely. The solenoids on her stomach had seized up, which overheated the wires running from there to the computer in her bra. On undressing, she found that the wires had actually burned a hole in her chest. "When I saw the charred flesh," said Doyne, "I couldn't believe it. 'Ingrid,' I said, 'I want you to try, but not that hard. I don't want your burnt skin as an offering to roulette.'"

10

Sensitive Dependence on Initial Conditions

You can't know how happy
I am that we met,
I'm strangely attracted to you.

Cole Porter
"It's All Right with Me"

Robert Stetson Shaw, stoop-shouldered and bearded, looked something like Woody Allen impersonating Karl Marx. A physicist and founding member of the Chaos Cabal, he possessed talents in realms as diverse as gag writing and musical composition. "If I don't get my research grant this year," he once quipped, "I'm going to go home and live with my mother. Then they'll see what can be done with an abacus!"

When not living in a New Mexico commune or sleeping next to his computer in the physics laboratory, Shaw was a sometime resident at 707 Riverside, where he kept his piano stored in what came to be known as the music room. He would slip into the room at odd hours of the day and night, shut the door behind him, and play the piano for hours on end without pause. An appreciative audience would gather in the hallway to sit with their backs to the wall. They listened through the door to an amazing stream of sound: a nonstop, virtuoso performance of Bach (especially the Chromatic Fantasy and Fugue), Mozart, Scarlatti, and Shaw's own compositions, which incorporated styles ranging from classical to ragtime. He played from memory and composed sonatas with the ease of a jazz player feeling his way into a riff.

Doyne described Rob as "catalyst and seer of the Chaos Cabal," but Shaw owed his career as a physicist to nothing more calculated than being in the right place at the right time. He was a graduate student shuffling through life with his customary reticence when Bill Burke, a physics professor at the university, asked him to look at a strange set of differential equations. Burke knew that Rob had dragged an analog computer out of the basement of the physics building, and he also knew that this machine provided the perfect tool for looking at the behavior of differential equations.

What Rob saw on programming his computer with Burke's formulas sent shivers running up and down his spine. It was a moment of Archimedean discovery in which he sat staring at something completely new. Through iteration — a kind of mathematical stutter in which computers solve a problem over and over again — the machine had taken Burke's equations and tipped them from order into chaos. But this chaos had characteristics unlike the random behavior that physicists had previously known as chaos. First, it had been generated out of a simple system, and, second, the chaos itself displayed various kinds of internal order. Classical physics has always assumed that the complex behavior of randomness would only be described, if ever, by equally complex equations. But Shaw, on his first glimpse into the *terra incognita* of chaos, had discovered that the opposite was true. Chaos can be generated out of simple systems merely by iterating, or looping, them through the kind of regressive cycles that computers — and compulsive neurotics — never tire of making.

With Burke's equations programmed into his machine, Rob could actually draw pictures of this strangely ordered and deterministic chaos. Swirled onto the screen of a cathode-ray tube, the pictures looked variously like doughnuts or funnels or galaxies stretched out of shape. These diagrams of a world in which randomness and order coexist represent what are known in physics as strange attractors. Of the three basic kinds of attractor, the simplest is called the fixed point. Imagine a pan of water shaken so that waves roll over its surface. Stop shaking the pan and the waves dissipate, with the water eventually returning to a state of equilibrium. The water at rest has resumed what is known mathematically as its fixed point of attraction. The second kind of attractor, called the limit cycle, produces a regular motion repeated over and over again. Limit cycle attractors are found in waves rolling in and out along a coastline, or in the behavior of water rocking from side to side as it flows down a pipe.

Classical physics up to the present has been mystified by attractors more complex than those that are fixed or cyclical. If water in a pipe flowed in anything other than a smooth, or laminar, fashion, its behavior entered the previously inexplicable realm of turbulence. But the doughnuts and lopsided galaxies drawn by Shaw on his computer were pictures of turbulence, and in this case the turbulence was *not* random or mysterious or inexplicable. It had been generated out of a deterministic system, and it manifested its own forms of internal order. To explain the order that Shaw had discovered in chaos, one needs to understand the third basic kind of attractor, known as the strange attractor.

Think again of water flowing down a pipe. Now put an obstruction in the pipe. The water, if flowing slowly enough, will divide smoothly around the obstruction and meet again on the other side. Now place a drop of ink in the water. In the slow-moving current, the ink drop will pass around the obstruction and swing back to rejoin the central flow. But if you increase the water pressure in the pipe, at some critical velocity the flow lines on the other side of the obstruction, instead of rejoining each other, will start to wiggle. This wiggling is at first periodic, so that our ink drop flowing down the pipe will still exhibit predictable behavior. But by cranking up the velocity of the water still further, the drop will commence to behave chaotically, and it is by means of attractors — not fixed or cyclical, but strange — that one follows the chaotic ink drop as it tumbles down the pipe.

In programming his computer to look at Burke's equations, Rob had stumbled on his own equations of motion that penetrated, even if only slightly, into the unknown world of randomness. It was Claude Shannon who defined information as the amount of surprise you get on seeing something happen, and news spread fast around the physics department that Rob's equations were generating a tremendous amount of information. By iterating these equations — repeating them over and over again — Rob could watch the process by which simple systems tip from order into chaos. While following this progression, he noticed two salient facts, which one might call the laws of chaos. The first law posits the "sensitive dependence" of all systems on their initial conditions. The second law states that whatever differences exist in systems will tend to increase over time. In the language of chaos theory, this second law predicts the "rapid divergence of nearby trajectories." Given sensitive dependence on initial conditions and rapid divergence of nearby trajectories, one can expect small differences in systems to compound themselves over time into very large differences.

Without knowledge of chaotic solutions, and before the invention of computers capable of arriving at them, Poincaré described the essential insights of chaos theory as follows: "It may happen that small differences in the initial conditions produce very great ones in the final phenomena. A small error in the former will produce an enormous error in the latter. Prediction becomes impossible, and we have the fortuitous phenomenon."

As Doyne Farmer wrote in commenting on Poincaré's observation: "Modern computer technology allows us to simulate dynamical systems that produce the 'fortuitous phenomena' and take them apart to study how, when, and under what conditions sensitive dependence on initial conditions occurs." If the laws of chaos theory sound rather abstract, or of local interest to plumbers dealing with obstructions in their pipes, the larger implications of these laws should become evident when one mentions, for example, that sensitive dependence on initial conditions and rapid divergence of nearby trajectories could very well explain the development of species in Darwinian evolution.

Norman Packard said of his own surprise at finding strange attractors at work in the midst of chaos, "This idea of information generation is pretty powerful if you allow your imagination to wander a bit. We have dreams of generalizing the theory of information generation in chaotic systems to more general systems, like that of biological evolution. Two billion years ago there was a blob of pre-biotic chemicals on the earth. It mushed around to form a few strands of DNA and these reproduced and eventually formed life that got increasingly complex. And as it got increasingly complex, new information was generated. Each time evolution occurs, more information is being produced in these new, more complicated forms of life. Our hope is to quantify this kind of information generation in exactly the same way as we're now doing for chaos. One of the things we try to do, for instance, is predict *when* you're going to have chaotic behavior and *how* chaotic it will be. This chaos corresponds to the amount of information a system is generating. The more information it generates, the more chaotic it is.

The appeal in studying them is that strange attractors produce information with ramifications in all sorts of fields, ranging from the theory of evolution to ecology, sociology, economics, and the workings of the human brain."

Fascinated by chaos and the strange geometrical structures that govern it, Rob Shaw moved into his laboratory to work through the night on his computer. His advisers got worried. Having finished all the requirements for his doctorate, he was in the last stages of completing a dissertation on experimental superconductivity, a subject apparently unrelated to strange attractors. Rob had no more than two months of work remaining to finish the dissertation. His advisers came to talk to him. He could finish in one month, they said. When that didn't interest him, they whittled it down to a couple of weeks. But Rob wasn't listening. He was lost to chaos and there was no way anyone could bring him back to earth.

From the information generated by his computer, Shaw isolated different types of chaos and strange attractors. Many of his ideas were new, but others, he discovered later, had been arrived at independently by other scientists. Edward Lorenz, a meteorologist at MIT, had stumbled on the first strange attractor back in 1963. He was looking at models for weather prediction when he noticed something odd about convection currents. These display what might be called pockets of chaos — loops of information that demonstrate perfectly the laws of strange attraction: a sensitive dependence on initial conditions, and rapid divergence of nearby trajectories. The discovery of the Lorenz attractor, as this particular structure came to be known, had surprising ramifications in the everyday world, although its immediate effect was to explain why long-range weather prediction is impossible.

The Rössler attractor is named after another early explorer into chaos, Otto Rössler, who works as a theoretical chemist in Tubingen, Germany. Rössler is a gracious, soft-spoken man who lives surrounded by books, so that a conversation with him proceeds more in the nature of a colloquium. He pulls texts and citations off the shelf and piles them in front of him, while carrying on a kind of dialogue through the ages, with speaking parts given to Aristotle, Maxwell, Einstein, and anyone else who has something to contribute. He ascribes to Anaxagoras the earliest definition of chaos, but Rössler's own discoveries in the field are pedestrian, rather than literary, in origin.

He was walking down the street one day when he saw a group of children standing in front of a window. He joined them in staring at what proved to be taffy puller — a machine with two arms that stretch and fold a sheet of taffy over and over again. Rössler stood and watched the taffy puller for half an hour. He was utterly transfixed. Not by taffy making, but by the rythmic motion of the machine, which produced what he recognized as a perfect example of strange attraction. Rössler imagined two raisins placed in close proximity on the surface of the taffy. While the arms of the puller stretched and folded the sticky mass, he followed his two imaginary raisins along a path of successive iterations. They traveled away from each other in an eloquent demonstration of sensitive dependence on initial conditions and rapid divergence of nearby trajectories. Still standing in front of the candy shop, Rössler scribbled the equations describing the strange attractor that bears his name, although he personally prefers to call it the taffy puller attractor.

Rob Shaw's own discoveries in the realm of strange attraction were made public in equally quixotic fashion. Norman Packard was leafing through a copy of *Scientific American* when he found an advertisement announcing a prize being given by Louis Jacot, a French businessman, for the most original essay on the origin of the universe. Norman wrote for details about the competition, and then he convinced Rob to submit an essay on chaos. His paper, called "Strange Attractors, Chaotic Behavior, and Information Flow," was accompanied by a cover letter explaining the relevance of strange attractors to the theory of evolution. Shaw won honorable mention for the Prix Louis Jacot and an award of two thousand francs — about five hundred dollars at the time — which he spent on flying to Paris to pick up his prize.

This marked the first public awareness of work being done by the Chaos Cabal. But soon there would be a flood of interest in the nighttime discoveries made by this group of hackers and would-be gamblers. When reporters from *Newsweek* and the *Los Angeles Times* came to inquire about what the Dynamical Systems Collective was finding out in the phosphorescent green sea of chaos, they discovered researchers holed up in a laboratory that looked like a submarine conning tower full of computers, terminals, plotters, printers, monitors, dials, gauges, and other paraphernalia needed for tracking strange attractors through the murky reaches of turbulence.

The formation of the Chaos Cabal in 1977 was itself a nice example of sensitive dependence on initial conditions. Like Shaw, the other cabalists had jettisoned careers in one of the more established branches of physics to join him in programming chaos into computers. Jim Crutchfield, who had been Rob's assistant when he was studying superconductivity, had no problem picking up the new language of strange attractors.

Norman's conversion was more complex. Phlegmatic and easygoing as he might be, he was the star of the physics department. He had zipped through the curriculum and passed his qualifying exams all in the first year — an unusual feat — and everyone expected him to do his doctoral research in statistical mechanics, which is one of the more gentlemanly pursuits in classical physics. The faculty was dismayed when he announced he was joining the Chaos Cabal, whose research looked to them like a kinky hybrid between philosophy and computer programming.

"I had made a name for myself in the department," said Norman. "They didn't see my truly lazy and sluggardly ways until the following summer when I went out to Las Vegas to play roulette. That gave them second thoughts about me, and then when I began working on chaos they had third and fourth thoughts. I fell into utter disrepute, from which I was only rehabilitated when we won our National Science Foundation grant."

Doyne, having dropped out of school for a year and a half to work on roulette, had long since given up the thought of becoming an astrophysicist. By the time he dropped back into school he was already a connoisseur of chaos. As he wrote in the acknowledgements to "Order Within Chaos," his doctoral dissertation: "Had Rob never heard of the Lorenz attractor, none of this [the formation of the Chaos Cabal] would ever have occurred. I would quite likely have gotten bored with physics and dropped out, and I would now be happily playing my harmonica for the hippies down on the mall. Instead, Rob planted the seeds of chaos in my brain, and here I am trying to be a respectable scientist. Such is sensitive dependence on initial conditions."

Except for its reliance on computers, the formal study of chaos has little to do with beating roulette. In fact, the "Santa Cruz school of nonpredictable physics," as Norman called it, was doing its best to subvert the classical Laplacian assumptions on which predictability in roulette is based. "In his deterministic, classical dynamics, Laplace said that if you give me the position and velocity of every particle in the universe, I can tell you exactly what it's going to do a million years from now. He was wrong," said Norman, in his calm, almost laconic tone of voice, "and the extent to which he was wrong has only been realized very recently."

The Chaos Cabal remarked the fallacy in Laplace's assumption when they found, as one would never expect in his classical physics, that very simple systems can evolve from order into chaos. "Originally," said Norman, "people thought the reason behavior looked complicated was because only a complicated set of equations could describe it, equations involving many different interactions. It was natural for Laplace to think this. But it turns out that you can have complicated behavior even for very simple systems, systems with very few interacting components.

"In the case of water flowing down a pipe periodically, if you take your fist and knock the pipe, the water will jiggle around. It changes its motion as it jiggles. But then in time it settles down again. That's what it means to be an attractor. You perturb the system by giving it a knock. Its behavior is altered. And then it settles back to what it was doing before. But then again — and this is the philosophically interesting part — you may knock the pipe and the water, rather than resuming its periodic flow, may go off and do something completely different and *never come back* to what it was doing before. If that's the case, its previous state wasn't caused by an attractor; it was unstable.

"These arguments against the Laplacian world view are disconcerting to a lot of people who thought the world was predictable in principle. And the fact that it *isn't* predictable has all sorts of philosophical implications that have yet to be worked out. For example, take the ongoing debate about determinism and free will. The determinists use the Laplacian world view to argue that the motion of physical systems — humans included — is determined in advance by the laws of physics. If this is the case, then there is no sense talking about free will. Your motion for the rest of time is already programmed. How strange attractors affect this argument is not trivial, and they do affect it, because strange attractors allow for the possibility of *spontaneous* change, so that your behavior for the rest of time can in no way be determined."

The revolution currently taking place in the midst of physics is one whose battles are being fought on the insides of computers. They alone — never tiring in their electronic compulsion to repeat — are capable of the iterations required for tipping simple systems from order into chaos. Only through the doggedness of silicon can one look at sensitive dependence on initial conditions and project this dependence far enough into the future to watch for divergence of nearby trajectories. Taffy pullers have been put to good use, but the new physics owes its insights and methods to the electronic computer.

"It's only with the advent of modern digital and analog computers," said Tom Ingerson, "that people have been able to solve complex mathematical equations in all their glorious horror. Instead of studying the real forms that equations happen to have, physicists for years tried to beat them into forms which were familiar. Some curves had names and others didn't. But Mother Nature didn't give a damn for these distinctions, and she went on being as complex as ever. Computers don't give a damn either. They can treat equations however they happen to behave. This field of strange attractors and chaos, or nonlinear dynamics, or whatever you want to call it, has arisen because people are attempting to understand the mathematics associated with more general forms of equations, which don't necessarily describe particularly complicated phenomena. It's just that Mother Nature doesn't happen to be linear all the time. In fact, she seldom is. One of the fascinating things that's come out of the study of chaos is the fact that a large class of nature is unpredictable. This still blows some people away when you explain it to them. But there are certain equations that indicate the *intrinsic* unpredictability of systems, and that's truly an amazing fact."

The Chaos Cabalists, in their published papers, and later when lecturing at scientific conferences, became known as artists in the realm of silicon. Out of their laboratory full of computers and printers came some of the earliest pictures of these strange worlds that displayed, as Doyne put it, a "peaceful coexistence between order and chaos." As preparation for launching themselves into the forefront of theoretical physics, Doyne and Norman had spent a year and a half at home building roulette computers from scratch. Rob Shaw had resurrected an analog computer from the junkheap of outmoded technology, and Jim Crutchfield had elevated the profession of computer hacker into a way of life. Their exaltation had humble origins, but the Chaos Cabal had been in the right place at the right time, with the right technology.

"These guys are real lucky," said Ingerson. "They got in on the ground floor, with the perfect combination of factors for success in science: native intelligence, opportunity, luck, and resources."

When its theories captured the attention of *Scientific American*, Douglas Hofstadter wrote in a review article on the field of strange attractors and chaos that "the simplicity of the underlying ideas gives them an elegance that in my opinion rivals that of some of the best of classical mathematics. Indeed, there is an 18th- or 19th-century flavor to some of this work that is refreshingly concrete in this era of staggering abstraction.

"Probably the main reason these ideas are only now being discovered is that the style of exploration is entirely modern: it is a kind of experimental mathematics, in which the digital computer plays the role of Magellan's ship, the astronomer's telescope and the physicist's accelerator. Just as ships, telescopes and accelerators must be ever larger, more powerful and more expensive in order to probe ever more hidden regions of nature, so one would need computers of ever greater size, speed and accuracy in order to explore the remoter regions of mathematical space. By the same token, just as there was a golden era of exploration by ship and of discoveries made with telescopes and accelerators, characterized by a peak in the ratio of new discoveries uncovered to money spent, so one would expect there to be a golden era in the experimental mathematics of these models of chaos. Perhaps this era has already occurred, or perhaps it is occurring right now."

While sailing their computers through the golden age of chaotic exploration, the Cabal never forgot the port from whence they had embarked. Doyne had "chaos in my brain," as he put it. But he was equally obsessed by roulette. New fields in physics do not remain new for long. One either seizes the moment of discovery or watches it slip into more enterprising hands. But the same was true of beating roulette. For all the Projectors knew, there might be dozens of engineers in the Silicon Valley tinkering in their garages with roulette computers. The Chaos Cabal had flags to plant on unexplored terrain. Eudaemonic Enterprises had a Pie to fill with roulette winnings. Their only choice was to do everything at once: roulette by day and chaos by night. Or vice versa.

"The Project gave us a weird perspective on the academic experience," said Norman. "Whenever we approached the end of the quarter or vacations, there was not only the pressure of finishing courses and writing papers and giving lectures, but there was also the pressure of getting the computers ready to go for one of the trips scheduled during vacations. We tried, but we never really had time to do everything."

The Projectors developed a manic schedule of work and more work. Every break from school they crammed computers and roulette gear into the Blue Bus and shot over the mountains to Tahoe or Reno or farther across the desert to Las Vegas. They became quick-change artists adept at transforming themselves from graduate students into gamblers. They learned fast how to wipe every sign of intelligence off their faces. One day they'd be programming a PDP 11/45 mainframe computer with Belousov-Zhabotinsky reactions or Lyapunov exponents, and the next day they'd be checked into a flophouse off the Strip practicing betting patterns with computers strapped to their chests.

Following the summer campaign in Las Vegas, the Projectors took the earliest opportunity to head back to Nevada during Christmas vacation. They were still engaged in the process, as Doyne put it, of "exploring the envelope."

For the Project's Christmas trip, packed into the Blue Bus alongside the computers and roulette wheel were Doyne Farmer, ace data taker, Ingrid Hoermann, veteran player, and two people new to the Eudaemonic adventure — at least the traveling part of it. John Loomis was back to try out the solenoids he'd redesigned the previous spring for the Reno trip. Living in Project Artaud, an artists' commune housed in an old factory on San Francisco Bay, Loomis was still supporting himself as a carpenter. "John worked long stints with breaks in between," said Doyne, "which was just the right kind of schedule needed for doing the Project. He was a real trooper, and he turned out to be a good person to have along." The fourth passenger riding the Blue Bus to Las Vegas was Neville Pauli, a former classmate of Letty's from Stanford Law School who was being groomed as the Project's high-stakes bettor.

On reaching Las Vegas on December 12, the Projectors followed their usual routine. They turned off the Strip into one of the seedier parts of town and started knocking on doors. They knew all the motels that catered to transients and gamblers, no questions asked, just pay in advance, first, last, and deposit. The trick in these joints was to front as little money as possible, because the Projectors were savvy by then to the fact that money in Las Vegas seldom makes its way back out through the cashier's grill.

The troupe checked into a two-room suite at the Brooks Motel, a mom-and-pop operation just off the Strip on Paradise Road. They pulled the blinds and unpacked the computers and roulette wheel. Neville was jumpy. He was sure that outside on the deck next to the pool people could hear the roulette ball spinning on the track. It didn't matter that it was December and that no one was outside. He insisted they keep the TV turned on to mask the sound. "We watched a lot of Sesame Street that trip," said Ingrid.

After setting up the biofeedback machine, John worked on training himself as a data taker, while Neville and Ingrid practiced flipping chips onto the Project's mock layout. "Neville developed a special costume for playing in the casinos," said Doyne, "a kind of Kiwanis Club conference look with a blue blazer and short haircut. He looked very intense, like a rich dentist or something."

Even though all the connectors had been newly insulated and the grounding system improved, the Projectors suffered their usual hardware problems with short circuits and loose wires. Fewer shocks came from the antenna T-shirts, but the players encountered a new problem caused by the winter temperatures. Moving in and out of casinos, passing from cool night air to overheated rooms, they found the frequencies on their radio receivers going off value. This problem with "thermal drift," as Doyne called it, was complicated by the fact that the Bus had no heater. So they had to drive around town keeping the computers warm under sweaters placed on top of the engine cover.

Ingrid wore her rabbit fur coat and tried to look as seductive as she could with a computer and battery packs stuffed in her bra. Doyne wore knit pants with a polyester shirt and ski jacket. "John and I looked like your average hicks," he said. Pauli fussed over getting the final touches right on his Kiwanis Club attire. Teamed up with Doyne for his first session at the Hilton Hotel, Neville in a few minutes cleared three hundred fifty dollars. When they rendezvoused later back on Paradise Road, Doyne was surprised to find him furious. There had been too many mix-ups with the signals, he said, while from Doyne's perspective, there were no more than should be expected from an initial playing session.

Keeping their bets low, playing one- or two-dollar stakes, the Projectors inched the bank ahead seven or eight hundred dollars. They wanted a cushion under them before pushing the stakes. Even with their massive advantage over the house, they knew that statistical fluctuations could bury them deep in a hole before the law of large numbers arrived for the rescue. They tried to log three or four hours of play a day, and typical entries in the Black Book show average bets as low as sixty cents.

They favored a few casinos with good wheels: the Riviera, the Silverbird, and Caesars Palace. But after another big win at the Hilton, they ventured farther down the Strip to the MGM Grand — a barn of an establishment with a gaming room that looks like a converted supertanker. The wheels at the MGM were spinning fast, but they were also nicely tilted; so Doyne bought into a game, adjusted parameters on the computer, and signaled for Ingrid to start betting. Playing more than two hundred trials on the graveyard shift, Ingrid at the end of the session had lost four hundred forty dollars. "When I found out how badly we had done," said Doyne, "it felt to me like statistical fluctuations. I thought conditions were good, and that's why I'd hung in there so long."

"The table was really crowded," Ingrid remembered. "I couldn't get a place near Doyne, and I kept losing his signal. We had decided to bet higher than normal stakes; so we were pushing money out faster than usual. I went through a hundred-dollar stack of chips, and then another and another and another. But it's funny. Because I was losing, I wasn't paranoid at all.

"The MGM has these underground corridors with spongy floors that run from the casino to the street, and we felt incredibly depressed walking out underground. It was worse in a place like this, where everyone was dressed up and slick. It had something to do with it being so anonymous, like a department store, except for the cocktail waitresses parading around in little costumes. It's a middle-class fantasy with nothing to it. We walked out feeling like kids who had tried to play a prank and couldn't pull it off. Surrounded by so much money and so many rich people, we couldn't even begin to make a dent, and in the end they'd get us anyway."

On returning to the motel Doyne and Ingrid found Neville in an agitated state. They were late. Why hadn't they phoned? He thought they'd been roughed up and dumped in Lake Mead. Everyone was tired, strung out, discouraged. John had been planning to leave the following day, and Neville joined him on the plane to San Francisco. Doyne and Ingrid checked out of the Brooks and packed the wheel and computers into the Bus.

Doyne headed for the highway, but Ingrid insisted they stop downtown for one last look at the wheels. "We parked on a street behind the Fremont casinos," she said. "Doyne was depressed after our session at the MGM. We sat on the grass in front of the Court House and stared at people revving their engines. Then we split up and walked through the casinos looking for one last wheel to play.

"When we found each other later, Doyne told me he'd spotted a good wheel at Sam Boyd's California Club. This is a workingman's casino, one of the smaller and sleazier of the downtown joints. It was Friday night and really crowded with people who had got off work to go gambling. There were only two wheels in play, far back in the club. Doyne set his parameters and gave me the signal to buy into the game. I went into the bathroom and turned on the equipment."

"There were a lot of funny people in there that night," said Doyne, "old ladies and sheepherders, a guy playing with his daughter, and a German immigrant. The German invited Ingrid over to his house for Christmas dinner, while I struck up a conversation with the pit boss, who was a twenty-eight-year-old blonde. 'How do you like being a pit boss?' I asked. 'How's life in Las Vegas?' It was weird and a bit distracting, because I felt as if I could have asked her out for a date. Ingrid and I played for three hours, and it must have been one of the most relaxed sessions we ever had.

"Our bank kept going up and down, up and down, but it went

up steadily between all the downs. As we sat there, I watched three-hundred-dollar swings in either direction. It was one of those frustrating sessions where the ball keeps falling just to the right or left of the predicted cups. Or you bet on numbers thirty and nine, and the ball lands on number twenty-six, the pocket between them. We'd raise stakes and lose, and then lower them and win. All in all, I didn't think we were doing very well. But I make a point when I'm data taker of not paying too much attention to the bettor and how many chips she has. It's irrelevant how much money she's making. I'm only worried whether the ball lands where the computer predicts it will."

"I actually happened to be winning a lot of money," said Ingrid, "and I didn't know whether I should act surprised. It was so consistent that it seemed hard to get excited about, especially since I knew I was going to keep winning. If I screamed or giggled or clapped, what was I going to do the next time the computer won, or the time after that? This could get real old after three hours, and I didn't want to draw any attention to myself. But the guy who had invited me to dinner was watching me and getting excited and exclaiming, 'God, you won again!'

"'Some days you win,' I'd say, trying to act casual, 'and some days you don't.'

"I was tired and spaced out after a couple of hours of playing, and I missed parts of the betting pattern that often held the winning number. But by that point we were making so much money I didn't get upset about it. Things were going so well that I just kept raising the stakes. By hiding my five-dollar chips under lesser denominations, not even the croupier knew how much we were winning until the end, when I scooped up all my chips and cashed out."

On meeting Doyne later at the Blue Bus, Ingrid reported clearing more than a thousand dollars. "It was a psychological shot in the arm," said Doyne. "The Project might have been dead without it. The money didn't matter so much as the fact that the computer had done exactly what it was supposed to do. I would have jumped up and down with joy, if I hadn't been so exhausted."

II

Small Is Beautiful

It is surely a great calamity for a human being to have no obsessions.

Robert Bly

Within the space of thirteen months, the Project had made no fewer than eight forays into Nevada. Beginning with Doyne and Alix's shadowboxing at South Lake Tahoe and Reno in December 1977, there had followed in quick succession the New Year's trip to Las Vegas, with Doyne's first "big" win at the Golden Gate Casino, the spring training session at Ralph's "cabin," the three summer waves to Las Vegas, the recent Christmas trip, and a January 1979 junket to Reno made by Norman, Jim Crutchfield, and Jack Biles. Clearing several thousand dollars in profit while betting mainly dimes and quarters, they had proved the system worked. But the Project, if it was going to continue, needed to upgrade itself from bush to major league. The Eudaemons needed a higher rate of return on their investment of time and money, which called for a larger pool of betting capital, more reliable equipment, and better training. "In the gambling business," said Ralph Abraham, "either you're a professional or you're nothing."

"We had been bootstrapping the bank up from zero dollars," said Doyne, "by betting only what we made in 'profit.' This was a good conservative strategy to follow while we were testing the system, but we were overly cautious. We should have switched gears and pushed the stakes harder a lot earlier. It's easy to go into a casino and get nervous and play smaller stakes, because you figure that way you can't hurt anything, or get hurt. But you're not taking into account what this does to you psychologically. Over the long haul it wears you down. Playing dimes and winning fifty dollars with the computer, you know in principle you've done something great. But it's not nearly as impressive as playing ten-dollar chips and winning five thousand dollars."

The Projectors had been plagued for a year by hardware problems — loose wires, bad connections, shocks, clamping solenoids, drifting signals. If they were going to transform themselves into professional gamblers, they could no longer tolerate having wires draped all over their bodies. Out to the scrap heap would go the first generation of Eudaemonic computers, along with the sacroiliac belts, solenoid plates, antenna T-shirts, and computer bras. The Project at this advanced stage required a new generation of computers — more compact, reliable, and efficient, with greater integration and fewer chips. "If you can get your chip count down to one," said Ingerson, "then you've solved the problem, at least internally, of bad connections." The name of the game for Eudaemonic Enterprises, and everyone else in the computer business, was miniaturization. "Small is beautiful" had become the clarion call for both micro- and macroeconomics.

They thought it might take six months to build a new generation of equipment. But as Doyne and the Chaos Cabal got deeper into the mysteries of strange attraction, they found themselves with less and less time for the Project. Having already spent a year and a half out of school building roulette computers, Doyne was now back at the university with papers to publish and classes to teach. To "professionalize" the equipment would require another large dose of time and talent. The computer needed to be redesigned, reprogrammed, and rebuilt from the ground up. Who would be crazy enough to tackle an assignment like this? It wasn't the kind of thing you could put on your résumé, and Eudaemonic Enterprises didn't have much to offer by way of cash incentives. A slice of Eudaemonic Pie delivered at some unknown future date sounded appetizing enough, but anyone smart enough to build a roulette computer could go to work at Intel with a starting salary of $35,000 a year, not to mention stock options and other sweeteners. The Project was on the verge of expiring after a few tantalizing successes. Its only hope of surviving lay in the unlikely event that it fell into the hands of a hacker possessed of knowledge in computers, electronics, physics, mathematics, and information theory. Furthermore, this hacker would need to be unemployed, close-mouthed, good with his hands, and free to work nights, weekends, vacations, and every moment in between. In exchange for signing on with Eudaemonic Enterprises, said hacker would receive a minimum wage, paid irregularly, and a slice of Eudaemonic Pie. There might be only one place in the world where an advertisement such as this would stand a chance of netting even a single applicant. Along the mountainous fringes of the Silicon Valley, the Project put out word through the grapevine: they were looking for a hacker *extraordinaire*.

"I had the naive vision when we started the Project," said Doyne, "that we were all embarking on the same enterprise. We were going to be partners. We were going to take it seriously. We were going to stick to it until it was done. And because I thought we were all in it together, there were times I resented being the one who had to go out on a limb. I didn't imagine I'd be a more significant participant than other people, but I ended up doing a huge share of the work by myself. I guess I really wanted something like the Project to happen. I viewed it as my chance to break out of the pack, and when I get determined to do something, I don't give up very easily. But at this stage, unless someone else took the initiative, the race wasn't going to get run."

As they thought of ways to save the Project, one idea kept reappearing as the perfect solution. In order to get all the hardware off their bodies, what if Eudaemonic Enterprises, as if by magic, could compress their computer, batteries, antennas, and solenoids into a shoe, or, at most, two shoes? Was it possible to make an ambulatory computer? Could one really fit the circuitry of a computer *and* all of its peripheral devices into a shoe? The answer to these questions lay in a feat of miniaturization not even attempted by the Japanese. Eudaemonic Enterprises had already made breakthroughs in the physics of roulette — by solving the equations of motion that govern the game — but their next challenge lay in building a computer small enough to operate out of a shoe.

"We were convinced the Project was going to work," said Norman, "but by the end of the January trip, we realized we had to have greater reliability, because unreliability was killing us. And to make the Project work, we needed to put the computer in a shoe, which was going to be expensive. We guessed it would take five thousand dollars to rebuild the system to that scale. I had no money. I was deep in the red for my college education, and there were times I had to borrow money from Doyne or Letty just to pay my share of the household expenses. In retrospect, it seems crazy to me that I would go off on this hare-brained scheme when I was competely in debt. Obviously, the pie-in-the-sky aspects of it led us on. We were youthful. We persevered. The Project had been scraping by up to that point on Doyne's money. But he'd reached the bottom of the barrel. We needed an outside investor, or the computer would never make it into a shoe. That's when Letty stepped forward to say she'd finance the new hardware and provide enough betting capital for us to push the stakes in Las Vegas. She'd been lending money to Doyne along the way, but this was her first official involvement in financing the Project per se."

In preparation for building the computer in a shoe, Doyne asked Jonathan Kanter to redesign the radio system by which the data taker and bettor communicated with each other. Having unbraided his dreadlocks and backed off from the more overt manifestations of Rastafarianism, Kanter was supporting himself by commuting to the Silicon Valley to sell ideas. On one successful trip over the hill he had sold an idea to a video editing company, and this in turn kept him commuting back in their direction to sell more ideas.

The radio link had been the Achilles' heel of the Project from the beginning. It was always on the fritz, flaking out, suffering from thermal drift, or being overrun by noise. Magnetic induction had proved undetectable by the casinos, but it was also at times undetectable by the players themselves. Kanter's assignment was to get the transmitters blasting out a clean, no-nonsense signal.

"You can think of them as a magnetic field wiggling around," Norman said of the transmitters and receivers. "You put a loop of wire in the field and that produces a voltage, which has to be amplified in order to turn it into a signal that the computer can detect. I knew how to do this. But the engineering involved in implementing these ideas is a nontrivial task. We had already changed the design two or three times, and when Jonathan suggested an alternate design, we decided to go with it.

"The major problem had to do with filtering the signal to get it out. This wouldn't have been difficult if it had been strong and solid, but we didn't want that kind of signal, because the casinos could pick it up. We wanted to operate, as they say in the technical jargon, *close to the noise.* But any time you're operating close to the noise, you're in danger of falling into it."

In his redesign of the radio link, Kanter kept the basic concept of transmitting signals by means of magnetic induction. These were digitally generated and received, like the new stereo systems that flash computer script onto their LED quartz-lock tuners, but Kanter devised additional filters for cleaning the signal of unwanted noise. "These provided a quantum leap in accuracy," he claimed, and for two weeks of work rebuilding the radio link, he was paid $600. This was Eudaemonic Enterprises' first cash outlay for its new generation of equipment.

After Kanter's involvement, the Project went nowhere. He returned to the Silicon Valley, and no other hacker competent to the task of rebuilding the computer stepped forward. The Eudaemonic Pie again looked as if it were levitating skyward for lack of substance. With the Project stalled, Kanter's new transmitters and receivers, along with the rest of the Eudaemonic computers and hardware, were thrown into cardboard boxes. These were stored in the basement of the Riverside house, where for the next eight months they would quietly gather mildew and rot.

On a sunny afternoon in April 1980, Norman was strolling out of Mellis Market with several loaves of French bread under his arm when he ran into Mark Truitt, a former student of his. Truitt, at age thirty-two, was finishing the second of his two college careers, this one in physics, while the first had been in art and sociology. The brightest student in his class, he could have landed any number of jobs in the Silicon Valley. But he wanted nothing to do with military work and bomb making, and he realized that most physicists, either directly or indirectly, do just that. The very afternoon he bumped into Norman, Truitt had walked out of a job interview at Watkins-Johnson, an electronics firm halfway between Santa Cruz and San Jose, on learning that they wanted his talents for building radar jammers for bombers.

"Working for the military is a job with no soul," he said.

"We have something you might like to do," Norman told him.

"Why don't you come over to dinner and we'll talk about it."

After dinner, Doyne, Norman, and Mark walked out of the house and down a flight of stairs to the basement. They unlocked the door to a small room with a cement floor and wooden shelves, and then opened a second door that led into the nether reaches of the cellar. Here, lying on a dirt floor, they found the cardboard boxes and suitcases in which the Project, unplayed now for a year and a half, had been stored.

"After explaining the system to me and asking if I was interested in building the next generation of computers," said Mark, "they thought I should look at the equipment. So we scrounged everything out of the basement and dragged it into the little room under the stairs. Out of these rotting boxes and suitcases they pulled a mess of antenna T-shirts, false-bottomed shoes, sacroiliac belts, battery packs, and computers. Everything stank of mildew. But we spread it out on the shelves and actually got one of the computers to do a little something.

"They gave me a battery pack and a computer to take home and study. I also had a shoe with a switch in it. I put them in my dresser drawer and took them out every once in a while to look at them. But all I could think was how unfeasible it was to put the computer in a shoe. Every time I looked at it, I realized what a giant problem it would be to fit the batteries in there, not to mention the rest of the hardware. I drew a picture of a large shoe with the batteries and computer in it, and the batteries alone, with several left over, covered the entire surface of the shoe.

"After making my diagram I stopped thinking about the Project. I just didn't see any way to take it seriously. I've always been intrigued by Robin Hood schemes, so I sympathized with the Project. But I didn't think it was the job for me."

Doyne had been invited to spend the summer and following year working on theories of chaos and turbulence at the University of Southern California's aerospace engineering department. Before moving south to Los Angeles, he arranged to meet Mark and talk about the Project at Banana Joe's, a student hangout at the university. Sitting in on the meeting was Jim Warner, the electronics technician attached to the UC Santa Cruz physics department. Warner knew about the Project and had already offered some advice on hardware problems. Doyne described to him the latest plan to put the computer into a shoe. Then Mark pulled out his diagram showing that not even the batteries could fit in that small a space, much less the computer, solenoids, and other circuitry.

"My God," said Warner, looking at the diagram. "What do you need so many batteries for?"

Doyne explained how the computer ran steadily over the course of several hours. The EPROMs, RAMs, transmitters, receivers, and solenoids required two different kinds of current — twelve and five volts of power — which in turn required two different bundles of batteries to keep everything running for a night of play.

"Why don't you turn the computer off?" said Warner. "When it's not making predictions, power it down. That would save you a lot of juice."

"We realized immediately that that was a good idea," said Mark. "It proved to be much harder to implement than we imagined. But this was the first time the thought of working on the Project sparked my interest."

Over the next few weeks Mark drew another diagram showing the circuitry required for turning the computer on and off. The microprocessor would run with full power when setting parameters or making predictions. At other times it would "power down" into a quiescent state that drew virtually no voltage. This greatly reduced the number of batteries required, and when Mark had finished making his second sketch, it looked for the first time as if the Project could actually build a computer in a shoe.

"Norman, Doyne, and I," said Mark, "talked about my diagram for the on-off circuit, and we agreed it was the way to go. But to build this kind of circuit into a computer requires a lot of extra chips, because when the computer powers itself down, it has to remember where it came from and how to get back there. You could say the new design was substituting memory for power. So when we looked at the on-off circuit and saw the extra chips needed to control it, we thought, 'My God, that pretty much fills the shoe right there.'"

Mark said he'd work on the problem. Again he threw the components into his dresser drawer, taking them out every once in a while to think about how to fit a computer into a shoe. He graduated from college that spring with highest honors and started work the next day as the sole employee of Eudaemonic Enterprises. He swept out the little room in the basement and built a workbench. He threw away the antenna T-shirts and aired out the computers. Surrounded by two oscilloscopes borrowed from the physics department, the KIM computer and PROM burner, the roulette wheel, a shelf of technical manuals, and dozens of plastic ice cream buckets filled with EPROMs, RAMs, and other electronic parts, he established the latest in a series of Eudaemonic laboratories.

Mark lived not far from the Riverside house, out the back gate and around the corner in a little bungalow that was once a vacation cottage. At odd hours of the day and night he could wander at will into the Shop, as he called it, to think about the three technical problems Doyne had given him to solve. Ingrid had been burned, and other people shocked, when the solenoids clamped open. At other times the circuit operating the solenoids had burned up completely. This was the first problem to fix. Then there was the on-off circuit to design, in which memory and logic chips would replace batteries. Mark's third and final assignment was to think of how exactly to fit the new computer into a shoe.

For such radical miniaturization, it was obvious that the Project's old technology of chips mounted into sockets and strung together with wire-wrap would have to be replaced by printed circuit boards — wafers of copper-coated plastic onto which silicon chips can be loaded directly without any intermediary sockets or wires. The actual construction of the PC boards would be subcontracted to a specialty house in the Valley. Doyne would handle any programming changes necessitated by the new design. But Mark would have to come up with at least a rough idea for the layout of the PC boards, as this was the only way he could gauge the final size of the new computer. For solving the solenoid problem, designing the on-off circuit, and making a mockup of a PC-mounted computer in a shoe, Mark would receive $2000 — payable on presentation of the first working shoe — as well as minimum wage for all his hours spent laboring on the Project. It looked to him as if he faced three straightforward problems, and he expected to have them finished by the end of the summer. Little did he or anyone else suspect that Mark would still be struggling with the details of building a computer in a shoe at the end of the *following* summer!

*

"Mark is all right brain," Doyne said by way of explaining Truitt's personality. Doyne was referring to the intuitive and associative powers, supposedly localized in the right hemisphere of the brain, that allowed Mark to freewheel his mind at liberty from one scientific insight to the next. But the fact that Mark was *all* right brain also entailed certain impediments. Missing, for instance, were those hemispheric sections specializing in language skills and the sequential processing of information. This meant that Mark's ideas, as brilliant as they might be, were not accessible through conversations conducted in the English language.

"Mark can describe something to me for hours on end," said Doyne, "and I won't have the slightest idea what he's talking about. I don't get the point until Norman comes along to translate for me. Mark's mind works by means of a filamentary process in which a half dozen ideas are all pursued simultaneously."

As Mark himself put it, "I have stray energy all over the place." His were the classic symptoms of hyperactivity. He was an intellectual speed freak, always *on*, whirling his mind from first principle to first principle with nothing assumed and everything doubted until proven otherwise. "I think the idea that light travels at a constant speed is merely a dogmatic assumption," he told me. "I tried to demonstrate this to a couple of my professors, and none of them would listen to what I had to say. It was too heretical. Norman alone among all the teachers at the university was willing to entertain the thought that the speed of light isn't constant. He was my teaching assistant in intermediate physics, and from that point on I visited him whenever I had other philosophical questions."

Doyne had also been Mark's teaching assistant for a workshop in electronics, where he was responsible for giving the class lectures on computers. "When we got to studying microprocessors," said Mark, "Doyne used his own experience with the Project as the basis for what he taught us. He actually brought one of the roulette computers to class one day, although I didn't know at the time what it was for.

"I'll never forget Doyne's first lecture on microprocessors. It was disorganized and gave me a headache and boggled everybody in the room, but I was intrigued. He was still talking half an hour after the class was supposed to have ended. We had a break for dinner and a lab that evening. Between Doyne's lecture and going back to the lab, I'd recovered and figured it all out. I flashed on the fundamental ideas needed to understand computers, and I haven't learned a whole lot new since then."

Of slender build, with brown eyes and a red beard, Truitt dresses, like the rest of the native population, in track shoes, blue jeans, T-shirt, and a digital watch worn prominently on the left wrist. But his springy walk and quick gestures are clearly those of someone metabolically notched higher than normal. There are moments in talking to him when you feel as if he might levitate off the floor in a burst of enthusiasm. He keeps interrupting and looping back on himself. Entire sentences are finished and dispensed with by the end of the first word. His elongated face, high cheek bones, bushy beard, and domed forehead give him the look of a young Dostoyevsky possessed alternately by visions of God and roulette.

Like most of the other Projectors, Truitt was a range-roving westerner who came of age with Sputnik and the war in Vietnam. Born in Santa Barbara, California, in 1948, the second son of a chemistry professor, he suffered what for his generation was the usual familial agon of high-speed motorcycle crashes, political revolt, and other torments deep in the Oedipal zone. "As a kid I was so hyped I gave other people headaches," he said. "If they'd invented Ritalin back then, they would have drugged me."

Growing up in a family of Christian fundamentalists, missionaries, and Quakers, Mark acquired their ethical concerns, purified them with existentialism, and decided as a logical consequence that revolution was the way to go. "I always took the most radical position. My family had a devil's advocate in me. I read Camus and Sartre, got deeply into existentialism, and ended up agreeing with Camus that if there is a God I'm against him. By the time I finished high school I was alienated from everything around me." Mark was also alienated from science and some of his own most formidable talents.

Attending a high school religious conference, he had been impressed by a debate he witnessed between a Quaker and a research scientist. "I thought the scientist didn't see the world around him. It seemed as if a scientific education involved you in too much detail, while it was more important to look at the larger picture. During my senior year I decided I didn't want to be a scientist. I quit taking math classes and tried to drop everything else. I thought I'd be a philosopher or a writer or a politician."

In 1967 Mark began the first of his college careers at Occidental College, with additional semesters spent at Swarthmore and in Mexico. "You can think of the chronology like this. I was at Swarthmore after the president died of a heart attack when students occupied the administration building, and I was in Mexico during the Chicago Convention. I was a full-time radical by then, busy occupying buildings all over the place." Nominally majoring in sociology, Mark was actually more interested in painting, sculpting, and writing short stories. After four years in college, he left without graduating. About to be drafted, he paid $500 to a Los Angeles law firm that specialized in keeping people out of the army. For added insurance, he made friends in Vancouver and other cities over the Canadian border.

Given the entelechy of the sixties revolution, Truitt drifted from politics into pastoralism. Working as caretaker on a fifteen-acre avocado ranch, he tended trees and did carpentry. He made his trips to Vancouver and put a lot of miles on cars and motorcycles shooting at high speeds up and down the Coast, until one too many brushes with mortality convinced him that his number was next. When Mark's best friend died in a car crash on Highway 5 at four in the morning as the two of them were driving from Berkeley to L.A., "it totally shook me up," he said. "I gave away my car and stopped driving after that. I used to feel immortal climbing mountains and doing reckless things. I'd always managed to luck out and get it together. But that was by far the heaviest experience in my life."

Rescuing him from the slough of despond, Mark's savior arrived in the form of a hometown Santa Barbara girl six years younger than he named Wendy Tanizaki. A Japanese American, Wendy possessed a luminous smile that shone out of a face animated with intelligence and some deeper kind of knowledge. Mark had dug a large organic garden. They planted corn together. When Wendy started college at UC Santa Barbara, Mark left his avocado ranch and followed her to the Coast, where he bluffed his way into a job restoring antiques. "Whatever work it was, I'd pretend to be an expert. But this wasn't an ordinary job. It was highly skilled work, more in the nature of doing a project or making art." Already a perfectionist, it was in Santa Barbara working as a furniture maker that Truitt polished his talents as an artist and craftsman — skills he would later apply to building computers into shoes.

When Wendy transferred to UC Santa Cruz, Mark followed her north and started the second of his two college careers. He enrolled as a freshman, majoring this time in environmental studies. "I was obsessed with solar energy," he said. "Then Wendy left for two years to live as an exchange student in Italy. I was completely bummed out, and as solace I started studying physics again. That's all I did day and night while she was gone." He took enough courses to get degrees in both math and physics. Recruiters from the Silicon Valley started tugging on his sleeve, and he began to think that "electronics might make an accessible career."

After a summer job building radio amplifiers, Mark had been offered a full-time position at the Stanford Linear Accelerator. But he had turned them down to marry Wendy on her return from Italy and finish his senior year at Santa Cruz. Only then, about to graduate first in his class in physics, did he look around the Silicon Valley for work. "Knowing that I could always go back to making furniture, I decided to give science another try. It couldn't hurt too much, and I might make some money."

On his initial sortie into the job market, Mark rode the bus from Santa Cruz to Watkins-Johnson — one of the electronics firms shoved, for lack of room, out of the Silicon Valley toward the coast. The Santa Cruz Mountains provide a natural barrier between the technologized interior and the once-placid shores of Monterey Bay, with their fog-shrouded redwood groves, grassy valleys, and fields of Brussels sprouts planted on bluffs overlooking the ocean. This was a Shangri-la favored by retired people of all ages, until the dominant culture made its first incursion into Santa Cruz by building a branch of the University of California. As the university came on line, producing new cadre for the job market, their commute inland got shorter every year as one high-tech division after another marched over the mountains to meet them on Highway 17. At last report, one of the Antonelli brothers had sold his begonia garden on the shores of the Pacific, and now the coast itself will fall to the advances of a chip factory.

Traveling five miles inland to Watkins-Johnson, Mark met a

classmate from Santa Cruz named Rob Lentz, who was waiting to be interviewed for the same job. They sat together through an orientation lecture, in which they learned that Watkins-Johnson wanted to hire them for a crash program building radar jammers for air force bombers. Disturbed by what he was hearing, Truitt left the interview. Lentz stayed, was offered the job, and took it.

"I walked out," said Mark, "because I knew if I had gone into the personal interview I would have attacked the guy. I had dropped science in high school because I thought the whole purpose of it was building weapons. I had finally overcome that idea, and the first job I apply for with my degree in physics, they want me to build radar jammers. I was shocked and disgusted. The electronics environment right now is incredibly militaristic. Let's face it, the government's going hog wild buying a lot of fancy electronic junk."

When Norman ran into him that afternoon in front of Mellis Market and invited him home to look at the Eudaemonic computer, it seemed to Mark that he had only two choices as a physicist. Work for the war department making bombs and bombers. Or work for Eudaemonic Enterprises beating roulette. The second, hands down, was for Truitt the more intriguing proposition.

12

Magic Shoes

Machines take me by surprise with great frequency.

Alan Turing

It requires strange talents to build computers. Given knowledge in math, electronics, and design, one also needs sympathy to bring inert pieces of silicon to life. You have to talk the language of machines — which is no more articulate than a series of electronic grunts telegraphed through a central processing unit in megahertz, or millionths of a second. You have to get down on the level of binary digits and coerce them two at a time through a maze of decisions. But while down there talking machine language, you must also comprehend the higher levels of computer thought, which is not really thought, but transistorized bits of silicon strung into loops of logic that only in the density of iteration *become* thought.

Once brought to life as thinking machines, computers, like children, have to be taught to pay attention. Computer programs are nothing more than attention-getting devices. The more complex the program, the longer the computer's attention span. It requires still more effort to teach these machines manual dexterity. Computers can perform a wide array of tasks. They can spot-weld fenders, fire ignitions, dial wake-up calls, buzz solenoids. But to push them from thinking about these assignments into performing even the simplest of them requires that they pay attention during the interation of *thousands* of logical steps.

To build a computer from the ground up, program it to play roulette, teach it how to buzz solenoids, give it the ability to transmit radio signals, mount it into a shoe, and then walk out the door wearing the world's first pedestrian model — this is a tall order. To fill it requires skills plucked from physics, mathematics, electronics, information theory, the fine arts, and shoemaking. Amazingly enough, Mark Truitt either possessed all these skills or was willing to bluff his way into acquiring them. With a history of rushing in where angels fear to tread, he was the perfect candidate for becoming the Leonardo da Vinci of computer design.

For all his scientific proficiency, the primary delight Truitt took in the Project was aesthetic. A computer in a shoe represented for him perfection in technological minimalism. It was the *mot juste* in silicon. He would go beyond canvas and pigments to make art out of printed circuit boards and chips. He believed the medium of microprocessors would produce the message of the twentieth century. It was in silicon that he thought people should look for the contemporary idea of beauty. The patrons of the new art were the Intels and Hewlett-Packards of Sunnyvale, who commissioned artists by the thousands to sketch diagrams of thinking machines, voice-activated machines, self-repairing and self-replicating machines.

Along with the temperament of an artist, Mark possessed what he called "natural rhythms," and if these dictated working only at night, or throwing a computer in a drawer to stare at it sideways while putting on his socks in the morning, everyone soon accepted his methods, because they worked. People got used to hearing him at odd hours open the gate into the back yard of the Riverside house, cross the garden, unlock the basement shop, and give himself over to dreaming about silicon cities built into shoes. Like any other artist attacking a major commission, he began by making sketches. He drew dozens of wiring diagrams for the on-off circuit and solenoid buzzers, wherein lay the technical problems he had to solve before starting work on the computer itself.

"I polished off the solenoid problem almost immediately, which was very encouraging. I discovered they were being driven off the computer by direct current. But the computer sometimes got lost in its program and went out of control, which in turn meant that the solenoid output went high. This drained all the current out of the batteries, melted the copper windings on the solenoids, blew out transistors, and burned the players. The solution was simple. I added a capacitor to the solenoid circuitry, which allowed only for pulses, rather than sustained drains of power off the computer."

Given his "filamentary method," Mark worked simultaneously on designing the on-off circuitry required to "power down" the computer when it wasn't setting parameters or making predictions. "I did the electronics troubleshooting early in the summer, and at the same time I worked out the idea for the on-off circuitry, not in wires, but the idea of it. I then got involved with the problem of fitting it all in a shoe, because I was worried that if I went ahead and built the circuitry, I still might not have enough room. It was at that point that I came up with the idea of a computer sandwich."

As far as Mark knew, no one had ever built a computer to be walked on. For a novel problem he came up with a unique solution. He would isolate the computer's two basic functions, logic and memory, into separate units, invert one of them, and snuggle it down on top of the other. One unit would operate exclusively in the realm of bits, the binary digits with which computers orient silicon into memory. It would hold the microprocessor, EPROM, and RAMs needed for operating the roulette algorithm. The other unit would hold a clock, five logic chips, and the transistors and amplifier by means of which the computer talked to the outside world through toe clicks, radio signals, and solenoid buzzes. "I had looked at all the components laid out on a two-dimensional surface, and I thought there were too many of them to fit into a shoe. That's when I hit on the idea of making two PC boards that would fit together like an Oreo cookie. There was going to be a tradeoff between thickness and overall size. But I was convinced the design would work." Using little more than a hundred dollars in electronic parts, Mark would eventually succeed in building an ambulatory computer sandwich that when finished was roughly two inches wide, four inches long, and a half inch thick.

"I pursued all these ideas in a scattered, or at least nonsequential, fashion. I was working on the design for the computer sandwich before I had even finished debugging the solenoids." Concerned that Mark might get lost in his invisible city, Doyne came up from Los Angeles at the end of the summer. In a week of nonstop work, the two of them cleared up the solenoid problem, finalized the design for the on-off circuitry, rewrote the computer program with instructions for handling the new circuit, and sketched the first complete pictures of a computer in a shoe.

"Toward the end of that first summer," said Mark, "I was working like a dog. I was completely obsessed with it." But a new problem had appeared that was sapping attention from his other filamentary concerns. Mark had retrieved Harry from among the old computers stored in the basement. He had dried the mildew out of it and set up the computer on a bench in the Shop. "I wanted to see it working perfectly before moving ahead to build the next generation of machines. But I discovered, when testing it, that Harry kept getting lost in its program. This was serious, because it resulted in the computer either giving out wrong signals or draining all the juice out of the batteries." Painstakingly checking every wire, socket, component, and chip for electrical continuity, Mark spent several weeks troubleshooting Harry's hardware. "I finally got the error rate down from three to one percent, which meant that every time you told it to do something, the computer would screw up one out of a hundred times. This might happen no more than once an hour, but it was still intolerable — no way can you let a computer get away with that sort of thing.

"I was incredibly frustrated. After everything I'd already put into the Project, I realized the computer still wasn't reliable enough. In the middle of September I said, 'To hell with this. I know I'm not supposed to mess with the program. But I'm suspicious, and I can't stand it anymore.' So I dragged the programming manual out of the basement and spent a week studying it. The manual by that time was a hundred and fifty pages long, and I went over it line by line. This was a last-ditch effort. I had tried everything electronic I could think of, and I was ready to give up."

At the end of the week Mark found the error. In the first line of a subroutine, a simple but necessary command had been omitted for telling the computer to pay attention. This was the kind of thing that only something as dumb as a machine would need to be told. Given the thousands of instructions that go into a computer program, the task of remembering what each of them means is simplified by gathering related instructions into electronic file folders that might be given names like "How to Drive Around the Mode Map" or "Roulette Wheels — Tilted."

"In programming," said Mark, "it's common to design subroutines made out of many separate modules. This saves space in the program. Instead of writing the same instructions ten times over if you want the computer to perform a certain task, you just zip around to the subroutine and back again to the program. I'm really into subroutines when I write programs. It saves a lot of effort when you have to perfect only one set of instructions for a particular assignment.

"The Project's roulette program is built out of subroutines. But there's one important trick to using them. When the computer goes to a subroutine, it has to keep track of where it came from. Otherwise, when it finishes, it has no way of remembering how to get back to where it was. Subroutines can use other subroutines which use other subroutines, and after you've nested them a hundred deep, you can end up with a very complex arrangement.

"Before branching into a subroutine, the computer is supposed to write the necessary information into its random-access memory. So the first line in the program should tell the computer: 'Remember the subroutine address of where you're going and where you came from.' It took me a week of staring at it before I realized this one command wasn't in the program. And because it wasn't there, when you turned the computer on, it would sometimes wander off to a random address and get lost."

Before starting work on actually constructing the new computer, Mark in another series of sketches designed a radio link to operate between the computer sandwich and its mode switch, which would be worn by the data taker in his left shoe. This second radio link eliminated all the wires that previously had run up and down the legs. In the new design, the data taker would wear a mode switch and transmitter in his left shoe, and a mode receiver, computer sandwich, and transmitter in the right shoe. The left shoe transmitter communicated between the mode switch and the computer, while the right shoe transmitter signaled the computer's predictions to *another* sandwich worn in the right shoe of the bettor.

On finishing his latest series of sketches, Mark was finally convinced that it was possible to build computer sandwiches — complete with microswitches, solenoids, radio transmitters, receivers, and batteries — small enough to fit into three magic shoes. He completed troubleshooting the solenoids. He finalized the design for the on-off circuitry. He satisfied himself that the program was debugged. He drew a sample layout of the printed circuit boards required for building the computer sandwiches. And then, when these tasks were finished, he went on strike. He locked the door to the Shop and stopped coming through the back gate.

"I quit working on the Project for a week. It was a crisis period." The crisis arose from the existence of that permeable, never completely definable boundary that exists between hardware and software. Except for Doyne's brief visit, Mark had been working alone throughout the summer as Eudaemonic Enterprises' sole employee. He had patiently sleuthed from one problem to the next until, finally, he was forced by necessity to track a number of glitches into the computer program itself. Programming the new computer was supposed to be Doyne's responsibility, while Mark was in charge of the design and construction of it: the hardware. This division of labor among computer builders is common but often impossible to maintain. Bugs scuttle from hardware to software through a barrier that for them is arbitrary and porous. While tracking them down, Mark himself had had to cross the line between hardware and software, and in doing so, a web of responsibility had settled over him for programming as well as designing and building the Project's computer. Consequently, he wanted two things, and he refused to go back to work without them. He wanted recognition for his new responsibilities, and money.

Late in October, with all the Projectors gathered in Santa Cruz for the annual Halloween party, Doyne, Norman, Letty, and Mark sat down to talk about how to get the Project back on track. "It had been such a frustrating time, and there had been so many difficulties," said Mark, "that I seriously considered dropping it. This was my first meeting with Letty, and she and the others responded to my concerns in a reasonable manner. Basically, everyone acknowledged that I had gotten involved with a lot of stuff that Doyne or Norman was supposed to do, or that was thought to be finished but really wasn't. So we spent a couple of days working out a new arrangement."

Part of this arrangement took the form of cash. The $2000 he had been promised was to be paid immediately. Mark's hourly investment in the Project, computed at minimum wage, had already mounted well over that figure. Added to this cash advance was a deferred salary, calculated at $10 an hour, to be paid from the Eudaemonic Pie. But the Pie itself at this point underwent a transformation. It was given a "front end," defined as the Project's earliest winnings, out of which Mark was to receive half of his deferred salary. The other half, and everyone else's share in the winnings, would come out of the old wedge-shaped, democratic Pie. For Doyne in particular, who had already accumulated a whopping thirty-five hundred hours working on the Project, the deal was hard to swallow. It compromised the basic principles of the Pie and valorized recent work over early work on the Project. But he had little choice about whether to accept the arrangement. Without a new generation of computers and another raid on the casinos, the Pie — with or without a front end — would never be served.

Everyone involved realized that the Project had become more complex than any one of them could control. Doyne was still master of a program nested in subroutines a hundred deep, while the arcana of the Project's transmitters and receivers remained in Norman's ken. Letty stood behind them as financial backer, and now Mark was recognized as CEO in charge of producing the computer itself. For the sake of Eudaemonic solidarity, they shook hands on the new contract and Mark went back to work. The strike was over, and the week's tension dissolved into Halloween revels.

Like its predecessors, this Halloween party had a theme. It was political in nature, a nod to the fact that communes and collective households everywhere were disbanding into more traditional nuclear entities. Conservatism was on the rise. A cowboy was being transposed live on TV from Hollywood to the White House. Money and power were back in fashion. Buttressing the new order was old religion — an aggressive form of Christian fundamentalism prepared to wage war on sin, the definition of which included most of what passed for daily life in Santa Cruz, California. So in the spirit of carnival, which embraces the demonic and purges it through dance, the Riverside communitarians invited everyone to come dressed on Halloween as "your favorite form of religious repression."

Riverside Street was flooded that night with a more luscious evocation of sin than appears in even the most hellfire of Jerry Falwell's sermons. At this black mass with a sense of humor, transvestite angels bebopped in the arms of red devils. The house itself had been transformed into various candle-lit shrines and grottoes. There were rooms devoted to mysticism, faith healing, the laying on of hands, and other rites both sacred and profane. One chamber held an Altar to Sex Roles made out of Burt Reynolds and Dinah Shore photographs, advertisements for bodily improvement, and the covers of Harlequin romances. In the dining room a cross had been constructed on which people could take turns being crucified.

Lorna Lyons came dressed as Saint Agnes, the fourth-century virgin martyred at age thirteen after rejecting the advances of a suitor. She wore a white toga, a halo of cotton balls, and enough makeup "so that I looked dead. Saint Agnes was a beautiful woman with men always on the make for her. But she refused to put out," said Lorna. The first man who tried to take her by force was struck blind and dumb. She took pity on him and returned his sight. But everyone was so enraged by her power that they cut off her breasts. I've always liked the fact that her withering glance could destroy."

Wearing a Green Beret uniform, reflector sunglasses, combat boots, and a T-shirt saying, "God, Mother, and Country," Jim Crutchfield attended as "a mercenary for Christ." Norman appeared as an apple tree with a snake in it. Ingrid arrived as a punk rock Mary. Letty came as Mrs. Money, with dollar bills pasted all over her. Doyne emerged as Mr. Self, the apotheosis of the "me generation." But late in the evening he was strapped to the cross in the dining room, where he underwent a symbolic crucifixion and resurrection. "I was converted into a non-me person," he said. "I was healed and lifted out of the misery of me-ness, so that once again I could see other people around me."

A few days before the party I had received a phone call from Doyne. "Are you interested in playing roulette?" he asked. "We have a trip coming up soon, and I think we could use you." Other than inviting me to play roulette with a computer in my shoe, Doyne had a favor to ask of me. He was going to be spending the year in Los Angeles at USC, and the rest of the Chaos Cabal was deep into their own research on strange attractors. This would leave Mark working alone on the Project down in the basement of the Riverside house. Would I mind dropping by the Shop now and then to say hello? Surrounded by RAMs, ROMs, and the twitter of a 150-page computer program stored in a tape cassette, Mark might appreciate the human contact.

Doyne and I were fellow students at UC Santa Cruz. This is primarily an undergraduate institution, with so few graduate students — about three hundred altogether — that it is not unusual for writers and physicists to talk to each other. I had known about the existence of Project Rosetta Stone for several years, but I was nonetheless surprised to learn how far advanced and well financed was the most recent of the Project's incarnations. Doyne phoned me at a moment when I was suffering from an advanced case of dissertationitis. This disease attacks graduate students in the terminal stages of writing their theses. The sickness, which can prove lethal, is compounded out of a mixture of ennui, insomnia, and other psychophysical symptoms ranging from hair loss to satyrism. No wonder I leapt at the chance to play roulette in Las Vegas with a computer in my shoe. This was a far cry from the burden of consciousness. Doyne assured me I would have no trouble getting good enough to work the system in my sleep. Of course there would be some danger involved. But it would be far easier, I thought at the time, to confront the Mafia than my dissertation committee. A pit boss might take me into the back room and ask a lot of tough questions. But when it got to that point, the trick in talking to the boys in Las Vegas would be to divulge as little information as possible. How refreshing! After being so forthcoming, responsive, verbal, and informed, I was delighted by the prospect of clamming up, going underground, stonewalling it. When it came to playing a part in this film noir, I assured Doyne I was a shady character he could count on.

The first opportunity I got to drop by the Riverside house, I found Mark out in the back yard pulling scraps of copper-covered fiber glass out of a deep-purple acid bath. "I'm making printed circuit boards," he told me. "I've never done it before so I thought I'd experiment first with small pieces."

If you want to make a computer from scratch, as Mark was in the process of doing, you go to the store and buy some logic and memory chips, a quartz clock, and a few transistors and other components that you then wire together into a circuit. Among the various ways to make these circuits, the simplest is wire-wrap. The chips get mounted into sockets with pinlike legs, plugged into a fiber glass board, and then wrapped with strands of conductive wire. The Project's early computers — Raymond, Harry, Patrick, Renata, and Cynthia — had been made by the wire-wrap method, but this technology was too bulky for producing something as dense as a computer sandwich. So Mark planned to switch to the more sophisticated technology of printed circuit boards.

A finished PC board consists of thin lines of copper foil laid over a fiber glass base. Into this map of lines, which represents an electronic circuit, computer chips are plugged directly without intermediary sockets or connecting wires. Hewlett-Packard and IBM, once their design departments have come up with a workable circuit, order their PC boards by the millions. Their production facilities are spiffier, but their methods for manufacturing PC boards are essentially the same as those employed by Mark in his backyard acid bath. First comes design. You gather together all the components needed for building an electronic city — its memory banks, data libraries, buses, and central processing units — and imagine linking them to each other in close proximity. Then comes layout. You tumble the chips onto a mock electronic terrain made out of Mylar, push them around, and figure out the best system of highways and access roads for connecting them into a working metropolis. Next comes artwork. You make a black and white street map of what the city will look like when finished, complete with dots representing the copper pads into which the chips will eventually be plugged. Then comes photo reduction and the process of masking the artwork onto the PC board. Finally, there is the etching of the board in acid, so that with most of its coppery surface eaten away, all that remains on the face of the board are the swirling lines of a metal circuit overlaid on a fiber glass ground. "When you reach this stage," said Mark, "all you do is plug things in."

Immediately after Halloween, Mark had run into what he called the PC problem. Very few computer manufacturers make their own boards. They subcontract the work to smaller companies that specialize in the design, layout, artwork, or photo-etching of printed circuits. Eudaemonic Enterprises had also planned to subcontract work on its PC boards, but on leafing through the Yellow Pages and phoning a dozen companies in the Valley, Mark was surprised to find that they charged more than $2000 for artwork alone, and another $500 for a prototype board. This was more than the Project had counted on spending; so Mark undertook to design and manufacture the boards himself. With the exception of the internal components, the Eudaemonic computer sandwiches would now be completely homemade. Mark was embarking on high tech as a cottage industry.

Using pencil and paper, he started designing a circuit. "It's like finding your way through a maze," he said. "You're working with a three-dimensional tangle of lines that you have to fit onto a two-dimensional surface. In the Valley there are entire layout departments that specialize in this stuff. They even have computers programmed to solve the maze problem for you. A computer can generate more versions of a circuit than a person can stomach. By the time I'd finished, I was absolutely sick of it, but a computer has no capacity to feel nausea."

Mark worked at a makeshift light table. Using tracing paper, he drew dozens of trial circuits for the components he had to stack and connect to each other. "Given the number of lines you're dealing with, you can't solve all the problems at the same time. So you do many versions of the circuit, and with each version you solve some problems and introduce new ones. You may get three lines properly connected, only to discover that another one is trapped. You can't cross lines over each other without their shorting out. So you work on the circuit one unit at a time, and when you get part of it looking all right you copy the lines onto tracing paper, before going back to fix another part that looks disastrous."

On solving the maze problem, Mark started the artwork. He used thin black tape to mask a prototype circuit onto a piece of Mylar that resembled a sturdier version of cellophane. He was working big, making a mask that would later be photographically reduced by half. Even on this larger scale, the tape was no bigger than .032 inches wide. After laying down the lines and making certain that none of them touched, Mark attached stick-ons representing the little doughnuts or pads of copper into which the chips would ultimately be pinned.

When he finished weaving a tidy maze of lines and pads, Mark took his artwork to a photo shop for reduction. They produced a negative of the layout exactly the same size as the prospective PC board, except that the black lines and pads in the original now appeared as transparent filigrees laid over a darkened background.

Transferring this negative onto a board is exactly like making a contact print in photography. Light is shined through the negative onto a surface that is then developed, washed, and fixed. But instead of using photographic paper, electronic circuits are "printed" onto copper-covered wafers of fiber glass one-sixteenth of an inch thick.

Mark had spent a month designing the circuit; now he had to manufacture the boards themselves. Working in a darkened room with light-sensitive lacquer, a 150-watt light bulb, a solvent bath, a sun lamp, and an etch solution of ferric chloride, he nursed his materials through the half dozen steps required to transform a negative mask into the positive tracings of a computer circuit. On pulling the first PC board out of its acid bath, Mark looked at the lines swirled on it and thought to himself that all the design problems involved in building a computer in a shoe had been solved.

"That day I was very high. As soon as I saw the first board come out of the bucket, I knew it was going to work." Drilling, loading, and soldering a circuit this small requires painstaking effort. The attached components would be hard to troubleshoot. Mark had yet to figure out exactly how he would fasten the boards together into sandwiches. And there was the final problem of building the computers into false-bottomed shoes. These "minor technical details" would take him another year of full-time work to solve, but as far as Mark was concerned, the Project was already a complete success.

13

The City of Computation

Uncertain fortune is thoroughly mastered by the equity of the calculation.

Blaise Pascal

The acacia trees are in bloom and the air sweet with the smell of freesias and other spring flowers when Doyne phones to ask if I want to go shopping with him in the Valley. He is not referring to food or clothing, but another item sold there at dozens of outlets large, small, discount, and deluxe. He is talking about shopping for silicon chips. The Project's homemade PC boards, after being drilled with holes and trimmed to size, are ready for loading with the hash of RAMs, ROMs, transistors, diodes, and other components out of which computers are made.

Leaving Santa Cruz early one afternoon, Doyne and I drive through the redwood groves along Highway 17 and over the Santa Cruz Mountains to San Jose. This megalopolis anchors the lower end of the Santa Clara Valley, now better known as the Silicon Valley. We turn north on Highway 101 and penetrate deeper into the pink smog of Santa Clara proper. Passing through the contiguous cities of Sunnyvale, Mountain View, and Palo Alto, we make stops along the way by exiting off the highway onto the wide boulevards of old farm towns that are now homogenized into bedroom communities and support services for Hewlett-Packard, Intel, Memorex, Teledyne, Synertek, Siliconix, and dozens of other companies with names assembled out of electronic acronyms. The computer factories themselves are housed in massive sheds built windowless for the sake of air conditioning. Other than tile roofs and glass entryways, they offer little more to the eye than acres of prestressed concrete. Surrounding the sheds are landscaped parking lots and fences with gates that control the flow of cars. Departing vehicles stop at the gates, while guards stationed in air-conditioned booths come out to search the briefcases of passengers, and sometimes their pockets.

"Theft is a big problem in the Valley," Doyne tells me. "A briefcase full of chips can net you thirty thousand dollars."

Rather than stopping at corporate headquarters, Doyne and I pull the Blue Bus into the unguarded parking lot at Halted, a surplus components store in Santa Clara that looks inside like the garage of a hacker gone wild. Cannibalized TV sets, radios, and antennas line the walls, while the rest of the building is filled with gray metal shelves overflowing to the ceiling with boxes of electronic parts. The contents of each box is marked in pen, with specifications given in ohms and farads, although many boxes are also labeled "Miscellaneous" or "?."

Doyne hands me a carton of "Assorted Resistors" and tells me to pick out the smallest of them, with values measured down in the range of mili-ohms. The cylindrical bodies of the resistors are color-coded in a rainbow of stripes, so that I hold in my palm what look like dozens of miniature African trade beads. Doyne gives me other boxes to sift through for capacitors measured in pico-farads, or 10^{-12} of a farad. Resembling tiny lollipops with paper handles, the capacitors also come color-coded in greens, blues, and purples. After buying several spools of hair-thin antenna wire, we walk out the door with $75 worth of merchandise slipped into a paper bag the size of a Mars Bar.

Driving north up the Valley, we pass what might be a storage park with dozens of cubicles for rent, until I see from the sign out front that this is a chip factory. We drive in front of another building constructed out of adobe arches strung next to each other in what looks like a mile-long McDonald's. After Halted, our next stop is Anchor Electronics, which turns out to be nothing more than a storeroom with a glass partition through which a woman dispenses chips. Doyne has phoned ahead, so the woman is waiting for us with a selection of CMOS RAM and EPROM chips packaged in two antistatic plastic tubes, each a foot and a half long. CMOS is the generic name for a family of chips designed to work at low power and under a wide range of temperatures and conditions. Good for cruise missiles and B-1 bombers, these chips also perform quite nicely in shoe-mounted roulette computers. Doyne needs more capacitors, so he stands at the window for twenty minutes running down a checklist of values. "We want the smallest you have," he repeats to the woman over and over again. She asks no questions, but soon stops supplying plastic bags for the capacitors. "At two cents apiece," she explains, "the bags are too expensive to keep handing out." Doyne pays her with a check from Eudaemonic Enterprises, and we walk out the door with two tubes of chips and a sack of capacitors for $187.

Trying to beat closing time, we push the Bus up the Valley to Zack Electronics in Palo Alto. Zack's looks like a hardware store with a long counter running the length of one wall. But the nuts and bolts being dispensed are of the high-tech, high-priced variety. Palo Alto is a classy neighborhood; so antenna wire that costs $2 at Halted goes here for $11.99. Doyne keeps two people busy behind the counter sifting through boxes of components. He buys a solder gun with an extra-fine point, a roll of antenna wire, four miniature 15-volt batteries, and a handful of resistors for $150. The last customers out the door, we filter through rush hour traffic and head back over the mountains to Santa Cruz.

Several weeks later, when I next see these chips and components, they are wired together into a prototype computer known as a breadboard. The word is also a verb, and the meticulous work of assembling computers is known as breadboarding. Norman walks into my house one afternoon carrying what looks like a box of Kodak photographic paper. "You might be interested in looking at this," he says, opening the box. "It's a breadboarded computer. We have here just about everything that goes into one of these gizmos." I look inside the box to see two pieces of white Styrofoam into which a handful of black chips and multicolored components have been pinned. They remind me of entomological displays in the Museum of Natural History, where one sees exotic insect species captured from microhabitats radically different from our own. Missing from Norman's collection are the name tags, but in their place are filigrees of wire running from specimen to specimen. "When the chips are finally loaded onto a PC board," says Norman, "they won't need any wires. The board itself will connect them into a circuit, and everything will be packed much more tightly together."

The components pinned to the two pieces of Styrofoam have been divided according to which half of the computer sandwich they will occupy. Segregated into memory chips and logic chips, the former species is the larger and more impressive of the two. The largest chip of all — obviously the queen of the collection — is a black-bodied rectangle with no fewer than forty legs. "That's a MOS Technology 6502 microprocessor," says Norman. "You might say it's the brains of the operation."

The microprocessor is the one component so important to the operation of a microcomputer that the words *microprocessor* and *microcomputer* have become almost synonymous. "There aren't a great many microprocessors to choose from," Mark later informed me. "There might be twenty different kinds, but most are part of a family of components. Of the five major families of microprocessors, each family speaks its own language. The largest family grew out of the 8080 made by Intel, which in turn developed into the Z 80 made by Zilog in Cupertino. This is one of the fifteen companies in the Valley owned by Exxon. The Eudaemonic microprocessor is part of what's known as the 6500 series. Originally made by MOS Technology, it got second-sourced by Mostek, a spinoff from Texas Instruments, which in turn got gobbled up by United Technologies."

Were its plastic or ceramic case cracked open, the black slug of a microprocessor, which is also known as the CPU, or central processing unit, of the computer, would reveal under a microscope a gray lattice of silicon. Wave after wave of this lattice is serried into the ranks of what are known as registers or store locations, some of which are devoted to arithmetical and control functions, while others are capable of memory.

"The main task of the microprocessor," says Norman, "is to shuffle data around, which it does through these forty gold pins. They're sticking in Styrofoam right now, but later they'll be soldered into a printed circuit. Each pin serves a different function, although the data running up and down it is limited to nothing more complicated than either a 1 or a 0. A pin can carry one bit of information, which means that it can be either 'on' or 'off.' But you can also put the bits together and shuffle 1's and 0's in chunks. A chunk of four 1's and 0's is called four bits. A chunk of eight is called eight bits, and so on. Each microprocessor has a characteristic word size, which is how many bits it can shuffle around in one fell swoop. Ours in an eight-bit microprocessor, which means that it manipulates eight-bit words. Other computers use words of different lengths, and word length is one of the main differences between microcomputers and larger machines. The IBM 360, for instance, uses sixty-four-bit words. A word that long requires huge amounts of processing, which is one of the reasons why those machines were never miniaturized."

Norman lifts the Styrofoam out of the box and points to several smaller chips pinned next to the microprocessor. "Your basic microcomputer consists of a microprocessor and a couple of extra memory chips. These two black ones are Harris 256-byte CMOS RAMs, which you can think of as the scratchpads of the system. Because they're RAMs, the computer can doodle in them while trying to solve equations, and then wipe them clean before tackling the next problem. These chips are new, low-powered, and very hard to find. Apparently not that many people care about saving a watt or two. CMOS, or complementary metal oxide semiconductor, is the name of a logic family that uses metal oxides as insulators and conductors. When you deposit these oxides on different layers of silicon, they form transistors. CMOS is only one of the several logic families. The most common is TTL, or transistor-transistor logic, while Hewlett-Packard uses SOS, which stands for silicon on sapphire."

Directing my attention to another chip pinned alongside the microproccessor, Norman fingers a purple rectangle that looks like a miniature California van with a sunroof on top. "This is the ROM. It stores the computer's program in long-term, or read-only, memory. This is a super deluxe model, a Texas Instruments 2532 EPROM with four thousand memory locations that can be reburned as many times as you want. It took three chips in the old generation of computers to do what this one does." He explains that the sunroof is actually a quartz window and that by shining ultraviolet light through it one can erase the electrical charges comprising the 1's and 0's of the computer's memory. The chip can then be reprogrammed by establishing a new set of charges.

In the hierarchy of computer memories, the simplest is fixed permanently in read-only memory (ROM). Up one level of computer intelligence is programmable read-only memory (PROM). But the most versatile of computer memories — which mimics what Freud described as the mystic writing pad of the human mind — is erasable programmable read-only memory (EPROM). This can be altered and augmented throughout the life of the chip. "Hackers joke among themselves about a fourth kind of memory known as a write-only memory, or WOM," says Norman. "You can put information into it, but you can't get it back out again."

On looking through the window of the EPROM, I see a gray mass of silicon sitting on a gold platter of conductive lines. "The lines along the edge of the chip decode and control its logic. The homogeneous gray expanse in the middle holds the four thousand memory locations. If you move your head back and forth over the window, you'll see rainbows. These come from extremely small etchings on the silicon that make up the locations themselves. Unlike the EPROM, if you opened a window into a microprocessor chip, its internal architecture would look far less uniform. You'd recognize discrete areas of silicon for performing different tasks. The complexity of what it does makes the structure of a microprocessor much more complicated than that of a memory chip.

"In fact," says Norman, "you could build a computer out of a microprocessor and no external RAMs or ROMs whatsoever. Its capabilities would be limited, because the memory chips store not only the information you're manipulating but also the instructions for operating the computer. But the microprocessor itself holds a half dozen memory locations, and this is sufficient to run the simplest of all computer programs, which says, 'Jump back to the instruction that says jump back to the instruction.' The microprocessor sits there jumping back to the same instruction over and over again in a little loop. As worthless as it sounds, we use this program quite a bit when we want the microprocessor to idle between clicks."

After showing me the EPROM, Norman places a piece of tape over its sunroof. "You have to keep the window covered because the program can be erased by sunlight. Ultraviolet rays break down the electrical contacts in the chip and reorient all the memory locations in the same direction. The program, which is stored in thousands of little charges built up between materials, simply leaks away. In that case we could always reprogram the chip, although that's sometimes trickier than it sounds. We've built a special circuit onto the KIM computer that addresses every memory location in the EPROM, pulses it electrically, and gives it whatever charge we want. But you have to be very careful not to apply the wrong voltage, or the entire chip goes up in smoke. We have a graveyard full of thirty-dollar EPROMs that have gotten fried."

Pinned into the Styrofoam next to the microprocessor, the two RAMs, and the EPROM is the black cartridge of a Synertek PIA, whose acronymous name stands for *p*eripheral *i*nterface *a*dapter. This is the fifth and final chip intended for the bottom half of the computer sandwich. "Your basic computer is made out of a microprocessor and memory chips," says Norman. "But if you want the computer to communicate with the outside world, you need some kind of interface."

As Norman explains its function, I get the idea that the PIA operates like the New York Port Authority of the computer. It provides the network and vehicles for moving bits by the million from chip to chip and farther out to the peripheral components. The PIA shuttles bits around the City of Computation on what are called buses. The buses come and go from data ports on the side of the PIA, and they travel through a grid of printed circuits or wires that stretch all the way from the central processing unit to the periphery of the computer. Riding on the buses are signals sent from the outside world into the CPU and back out again. On one side of the PIA are the blinking lights, buzzing solenoids, radio frequencies, key strokes, and phosphorescent words by means of which humans communicate with computers. On the other side is nothing but the silent shuttle of electrons pulsed through silicon gates at a million or more cycles per second.

Norman points into the Kodak box at a cluster of brightly colored wires blossoming from the data ports on the PIA. "Two kinds of buses arrive here," he says. "There are data buses and address buses. A bit decides to get on one bus or another according to the following schedule. When the microprocessor finishes calculating a roulette prediction, it needs to communicate the news to someone, namely the bettor. It does this by putting sixteen bits, or two bytes, out on the address bus and another byte on the data bus.

The bytes on the address bus arrive first and alert a particular pin on the PIA to expect a message. Then the data byte arrives to give it the message. This in turn tells the PIA to send a signal out from one of the data ports, and this final signal buzzes a relay, transmits a radio wave, or does any number of other useful things.

"The data bus is bidirectional, and the procedure works the same way for information coming into the computer. Let's say the microprocessor has to retrieve something from its memory. The memory, as you recall, holds four thousand bytes on a single chip. The microprocessor wants to grab *one* of these bytes; so it sends out an address on the address bus, which tells the memory chip which byte it wants. After the memory chip has decoded the address, it sends back the byte requested on the data bus."

Turning from the microprocessor, memory chips, and PIA, Norman lifts the second piece of Styrofoam out of the Kodak box. Pinned to it are the components destined for the top half of the computer sandwich. If the bottom is devoted primarily to memory, the top of the sandwich will specialize in logic. Five black rectangles, each roughly as long as a fingernail, are attached to the Styrofoam by their golden legs. "These are the logic chips," Norman announces. "They tell the microprocessor when to turn itself on and off. That happens to be a nontrivial task. Like everything else in the computer, the basic unit of information for the logic chip is a bit. A bit can be carried by one wire, which has either a low level, corresponding to a 0, or a high level, corresponding to a 1. Logical functions in a computer are performed by thousands of these wires built into circuits, and the circuits themselves are organized into various kinds of logical building blocks."

The "wires" and "circuits" in a logic chip are actually nothing more than microscopic locations — known as logic gates — etched by the thousands into the crystalline structure of a silicon chip. These miniature switching circuits have two states, open or closed, and exactly how these two possibilities allow binary numbers to solve logical problems is crucial to understanding how computers "think." The simple steps performed at the level of the logic gates provide the foundation for a computer's more advanced forms of cognition.

Computers begin their thinking process by shifting relationships between numbers into those between statements. The logic gates accomplish this task by translating combinations of 1's and 0's into "true" and "false." The mathematical logic of digital circuits — the rules by which this translation takes place — was developed by the Englishman George Boole, a nineteenth-century contemporary of Charles Babbage. These rules, known as Boolean algebra, allow for the expression of relationships that are simultaneously both mathematical and logical in nature.

The clearest explanation of Boolean algebra comes from watching the actual process by which a computer controls and operates its logic gates. Each gate consists of an electronic switch with two wires leading into it, and one wire leading out. A wire with voltage on it is said to be "high." A wire with no voltage on it is "low." The two wires and two voltages leading into the gate allow for four possible combinations. How the switch interprets these different combinations — and thus allows either a high or a low voltage out the other side of the gate — is controlled by something called a truth table.

A truth table might instruct a gate to open if and only if both wires leading into it are high. Or the truth table could tell a gate that one high wire is sufficient for opening. A switch operating according to the first kind of truth table is called an AND gate, while the second switch is known as an OR gate. The formal logic behind the working of an AND gate is expressed by the statement, "If, and only if, A is true *and* B is true, then their combination is also true." When the truth tables of AND and OR gates are inverted into negative propositions, they produce NAND and NOR gates, and these four different kinds of gates are the logical building blocks out of which all digital computers are constructed. The truth table for an AND gate is reproduced here. *A* and *b* represent the two lines lead-

operator	operands		result
	a	b	c
+	(↑ ,	↑) =	↑
+	(↑ ,	↓) =	↓
+	(↓ ,	↑) =	↓
+	(↓ ,	↓) =	↓

Figure 2. The truth table for an AND gate.

ing into the gate. *C* is the line leading out of it. An upward-pointing arrow represents a high voltage, or a 1, or a "true." An arrow pointing down equals a low voltage, or a 0, or a "false."

"You can specify what happens to the output line in a computer for every possible input," Norman informs me. "From these elementary logical devices, which are fairly easy to build electronically, you can make far more complicated systems. The microprocessor itself is composed entirely of these logical building blocks. So too is the memory. The microprocessor holds thousands of blocks, and each block in turn contains several transistors capable of being oriented as either 'on' or 'off.'

"Of the five logic chips used in the Project's computer, four are made entirely of NAND gates, which are AND gates inverted into their negative. The truth table for a NAND gate will have a 1 wherever the AND gate truth table has a 0. Before you go to the store to buy some chips, you first have to figure out what logical functions you want them to perform. Do you want AND gates or NAND gates? Once you decide on a function, you look in a data book to find out which chips are available for performing that function. Then you drive over to the Valley, step up to a counter, and say, 'Could I please have a 7400 or a CD 4001,' and they'll know exactly what you're talking about."

Late in the afternoon Norman declares, "That's it for the chips. The rest of what you see here are a few resistors and capacitors, an amplifier, a filter, and a couple of transmitters and receivers for sending signals from shoe to shoe. The only other really important part of the computer is this little aluminum canister. It holds the clock that governs the timing of events in the microprocessor. You need some sort of clock to step the computer through its sequence of activities. This one is made out of a quartz crystal that oscillates at one megahertz, which means that the clock puts out a little square wave that changes from 1 to 0 a million times a second. During every cycle of the clock, the microprocessor does something. It fetches an instruction, executes a command, or puts a byte out on the data bus.

"This gives you some idea of the speed with which all this is happening. It's going on very fast — much faster than human temporal perception — and this is one of the reasons why we use a microprocessor to predict roulette. Human beings are slow compared to the speed of electronics. We're doing quite well to get our reflexes down to a tenth of a second. That's a hundred thousand times slower than a microprocessor. Of course," says Norman, leaning back in his chair and grinning, "we can do other things that take a microprocessor a *long* time to do, like carry on a conversation."

Missing from this collection of chips and components pinned to the Styrofoam are a number of elements crucial to the working of the computer, such as batteries for running it and solenoids for outputting its predictions. But the brains of the operation — logic gates by the thousands oriented into a computer program — lie in front of me. "Basically, that's all there is to a microcomputer," Norman concludes, putting the top back on the Kodak box. "See how simple it is? One of the most amazing things about the computer industry is that you can use microprocessors and microcomputers — you can control them and tell them what to do and build them into circuits — without having the faintest idea what a transistor is, or what's actually going on electronically on a chip. Once you specify its characteristics, you don't have to know a single thing about how a chip is constructed. Even to first approximations, you can be totally ignorant about all this stuff, and you can be ignorant, as well, about the old-fashioned electronics. All you have to do is manipulate chips like little black boxes. It's magic, but all you care about is whether it works."

I climb the front steps of 707 Riverside one evening as a guest invited for dinner and roulette. I push open the front door, unlocked as usual, to find Norman sitting on the floor in the hallway. All six feet two inches of him is hunched over a white box that turns out to be, on closer inspection, a portable computer complete with keyboard, one-line display, and modem linking it by telephone to the mainframe up on campus. "I'm on line," he says, looking up to give me his crooked smile. "You know how it is. When the muse calls, you have to be there."

I leave him writing chaos functions into the machine and walk into the living room where Lorna, in black tights from her Jazzercise class, with her brown hair cut in the original Jane Fonda look (parted in the middle and falling around her face in terraces), is watching *I Walked with a Zombie* on TV. "It's sort of thirties French expressionism with incredible lighting," she says. "There are some *great* scenes." From her cockpit in front of the TV, Lorna runs most of the daily affairs at Riverside. If you want to know whether the kitchen is stocked with cumin, ask Lorna. If you want to find last month's phone bill, ask Lorna. She is newly re-enrolled at the university to finish her B.A., and coursework is cutting into her daytime viewing, but if you want to know the plot to any movie ever projected onto a screen, ask Lorna.

The smell of dinner cooking lures me through the dining room into the kitchen. Tonight at the stove I find Jim Crutchfield, which foreshadows Chinese cuisine of note. He is busy marshaling his ingredients on a long table in the middle of the room. I see piles of shaved vegetables, snow peas, bamboo shoots, curls of ginger, and slivers of fish marinating in rice wine. Assistant chef tonight is Grazia Peduzzi, a Milanese leftist now teaching Italian at the university. Dark skinned, with an aquiline nose and flashing smile, she breaks into gales of laughter as Crutchfield shows her how to milk the liquid out of tofu wrapped in cheesecloth. Flushed from an afternoon run on the beach, Doyne and Letty walk into the kitchen wearing cut-off shorts and jogging shoes. The last person to arrive, weighed down under a video camera and tape deck, is Ingrid. "I'm making a movie about women in science working together in groups," she says. "But it's taking forever. The funny thing is, I'm doing it with three other women and we can't agree on anything."

There is copious food for eight as the serving dishes fly around the dining room table. Doyne and Norman are legendary eaters from their days back in Silver City when they could pack away a couple of birds apiece at the Holiday Inn all-you-can-eat chicken night. On seeing that chopsticks do nothing to slow them down, Grazia breaks into another round of laughter. Crutchfield eats with the abstracted look of a chef tasting his sauces. Ingrid picks up a copy of *Good Times* and announces the movies playing in town. Lorna gives a running commentary on their actors, directors, and plots. She comes down hard on anything "sexist or sappy," but everyone finally agrees on seeing a double feature at the local rerun house of *The Damned* and *Death in Venice*. "That should be a heavy dose of freak-out," Ingrid says.

As we pass through second and third helpings, the conversation veers into talk about computers and evolution. "In fifty to a hundred years we'll be accustomed to seeing self-reproducing machines," Doyne declares.

"But at the moment we don't even have self-*repairing* machines," Ingrid says.

"It's not far from self-repairing to self-reproducing. Until they can reproduce themselves, computers aren't even as smart as amoebae. But from the first computer built in the 1940s to now is such a short period of time that their evolutionary prospects are amazing. From amoebae to slime mold to frogs to *Homo sapiens*, the next step up the evolutionary ladder might be machines."

"We could be the first species to design our own successor," Letty says. "The new species might come from one of two directions: genetic engineering or self-reproducing machines. But we don't even have the language yet to talk about these possibilities. Our words for machines and species and evolution are all too confined to the old meanings."

After dinner, as everyone else leaves for the movies, Doyne and I walk out the back door and down the stairs to the garden. We stand for a moment in the yard. The night air is sweet with the smell of magnolia blossoms and sour with the odor of anchovies washed ashore during their yearly run into the harbor. The earth in the garden is freshly turned. Over the back fence the horizon to the west is lit with opalescent corals and blues, as if ocean and sky are on the verge of exchanging places. Except for the swish of waves on the shore and the barking of sea lions swimming under the town wharf, the night is perfectly still.

We unlock the door under the stairs and walk into the Shop. The stairs themselves constitute part of the ceiling, and there are gaps in the walls sufficient for viewing the setting sun. The air is damp and acrid with the smell of solder. The workbenches and shelves lining the walls are piled high with electronic gear, including two oscilloscopes, the KIM computer, breadboarded circuits, and row upon row of plastic ice cream containers filled with batteries and components. Doyne tunes a radio to KFAT, the local country station, and Willie Nelson comes over the air singing a love song.

Most of the Shop floor is taken up with workbenches, a Heath power supply, a drill press, and a band saw. There is also a bookshelf holding a small library of reference works, lab books, microswitch catalogues, casino brochures, and novels. In between *The*

Radio Amateur's Handbook and a volume on transistor-to-transistor logic is Joan Didion's *White Album*. While most of the room is given over to Science, the western wall is reserved for Art. Hanging on it is a life-sized buffalo head fired out of clay and mounted on a trophy board. Surveying the room through red glass eyes, the buffalo wears a gold plaque under his chin whiskers that says "Manifest Destiny." Below the buffalo is a circle of chairs surrounding a table made out of cinder blocks and a wooden shipping crate. Stationed on top of the crate, and balanced on three stainless steel jacks, is the Project's B. C. Wills roulette wheel.

Doyne leans over the wheel and points to where the varnish has been burned by a stroboscope that exploded during the feasibility tests. He touches the chrome capstan and spins the rotor. Even under a bare light bulb, the red and black numbers meld into the hypnotic whirl of a roulette wheel in play.

But every few seconds the wheel goes *thunk, thunk*, until Doyne takes a piece of sandpaper and slips it into the tight spot where rotor and stator are rubbing against each other. "It's sad to see the wheel looking like this," he says. "We stored it for the winter under the stairs, and when I uncovered it in the spring I found a puddle of water sitting in the middle of it. The brass band around the edge of the rotor is badly warped. I should take the wheel back to Reno and get it refurbished, but for now we're making do with sandpaper."

When the rotor spins without rubbing against the housing, Doyne puts away the sandpaper and wipes the dust off the wheel and the mahogany rail that rises around it like the wall of a stadium. This venerable piece of wood, burnished through years of use, supports along its surface the parts to another game — a counterforce, an antiroulette — which will be played against the wheel in a contest of wits.

Not yet mounted onto a PC board, the Eudaemonic computer is still pinned to two beds of Styrofoam. Like the tendrils of a Day-Glo creeper, multicolored wires trail from the computer over the railing to several battery packs, two microswitches, and three solenoids. These make the system complete with a brain, power, input, and output. The switches, fastened onto pieces of wood, are worked with the index finger. Later, when built into shoes, they'll be operated by toe. The solenoids are mounted into a metal plate designed to be worn against the stomach, but they too will eventually fit into a shoe. Doyne plugs the batteries into the computer, and I hear the tinking of relays as it powers up. He sets the rotor spinning at an easy lope. As it revolves on its steel spindle without a sound, we're ready to play roulette.

With a practiced flick of the wrist Doyne launches a roulette ball onto the grooved track under the rail. It circles fast, passing in front of us a dozen times with the rattling sound of a coin rolling over a hardwood floor. It slows from friction and air resistance and then hangs for an instant on the lip of the track before surrendering finally to the pull of gravity. The descent toward one of the thirty-eight numbered pockets spinning below is charged with significance. While loose in space, the free-falling ball is open to chance, and as long as it struggles against entrapment on the central disk the game is still infused with hope.

The ball picks its way through the eight chromed diamonds dotting the stator. It whirls over the varnished bowl, strikes a fret separating two pockets, bounces and hangs for another instant before finally coming to rest. What was once an open field of possibility is now fixed into the determinacy of a number between 00 and 36. Over and over again, Doyne shoots the ball along the track. He taps the microswitches next to the computer with his left and right index fingers and watches the ball spin and fall.

"The switch on the left is for changing modes," he says. "I'm using it to drive around the mode map. The other switch is for entering data. It allows me to increment and decrement parameters."

He points above the wheel to Manifest Destiny. Taped to the buffalo's nose is a copy of the mode map, which is nothing more than a series of interconnecting loops, with a clicking code that allows for switching from one loop to another. The loops represent domains in the computer program, and each domain specializes in handling a different part of the roulette algorithm. This separation of the program into modes allows it to be fine-tuned to the unique characteristics, or parameters of motion, that vary from wheel to wheel.

"I've just driven into mode five," Doyne tells me. "That's the one for setting the rotor parameter. Like all the other modes, it has a preset standard value that has to be incremented or decremented according to what we find in the casinos. Tonight, for instance, the wheel is slowing down faster than usual. So I want to jack the parameter around to take account of an increase in friction.

"Conditions on a wheel can shift dramatically from one day to the next. The casino can change balls on you, or oil the wheel, or move it to vacuum the carpets. That's why it's important to hone the parameters just before playing. Otherwise your predictions are going to be skewed. You might find the computer telling you to bet in the shadow — that part of the rotor where the ball *almost never* lands, and in that case you're going to get creamed. Instead of a forty percent advantage, you're *losing* at that rate."

Doyne goes back to tapping microswitches and motoring the computer from mode to mode. I follow his path on the diagram taped to Manifest Destiny. As he steps through subroutines and adjusts parameters, the computer gets progressively smarter at tracking the game.

"It takes me about fifteen minutes to motor around the map, but it sometimes gets confusing. I get lost and can't figure out where I am, or the computer crashes. You can't pull out your mode map in the middle of a casino; so you have to have the instructions in your head. I've got to the point where I can cruise around the program without too many problems, but if I really get in trouble, I can do a total reset, which wipes out the parameters and dumps the computer back at the preset values."

With the practiced fingers of a pro, Doyne steers the computer through the eight subroutines in its program. He taps an ambidextrous code for entering, adjusting, and leaving modes. He calculates the deceleration of the rotor by clocking two of its revolutions past a fixed reference point. He does the same for the ball. At still another location in the program he calibrates the time at which the ball tends to leave the track. Five other modes establish variables for playing on flat, tilted, or slightly tilted wheels.

"It looks tonight like mode four is the ticket," Doyne says, on finishing his drive around the map. "Let's see how it does against the wheel."

He nudges the capstan and sets the rotor turning at a gentle clip. He launches the ball up on the track. Once the game is in play, every second counts toward beating its outcome with a prediction. Doyne clocks the green 00 on the rotor as it passes the reference point. He does the same for the ball. Four clicks . . . and the computer in *micro*seconds can play through to its conclusion a game that in real life takes ten to twenty *seconds* to enact. The computer calculates the coefficients of friction, wind resistance, and drag at work on the ball. It figures the rate of deceleration, position, and time of fall from orbit. It knows in advance the speed, distance, and path of the ball's trajectory down the sloping sides of the stator. It tracks the speed and relative position of the disk spinning below. It pinpoints one of eight octants of numbers spaced around the perimeter of the wheel and announces, several seconds in advance of its happening, where the ball will finally land on the revolving disk.

I hear the faint buzz of a solenoid coming from under Doyne's shirt. "Nine," he says, translating a high-frequency buzz on solenoid number three into a "no bet" signal. "I must have clicked too late."

Doyne again spins the ball on the track and clocks its progress over the varnished surface. Looking like a yogi meditating on a lower chakra, or a colitis victim suffering gas pains, he concentrates on the thumpers held to his stomach. "Six," he sings out, referring to a high-frequency pulse on solenoid number two. Octant six on the wheel holds the numbers 30, 26, 9, 28, and 0. Doyne and I wait as the ball slows on the track, hangs for a moment, drops from the lip, arcs down the bowl, skids among the frets, and come to rest in pocket number 9.

"It looks good," he says. "But you never know for sure how you're doing until you compile a success histogram with hundreds of trials. The only thing we're interested in is the law of large numbers."

We spin the wheel and ball through dozens of mock games. Doyne calls out the predictions pulsed onto his stomach, while I plot a histogram of data points recording whether the ball lands in front of, behind, or dead center in the range of predicted numbers. Doyne establishes a rhythm for spinning the wheel, launching the ball, clicking the switches, and reading the buzzes off his stomach. Spin, launch, click, buzz. His body is sprung tight with concentration. His reflexes are perfectly sensitized for translating the motion of wheel and ball into the digitized clicks of a computer program. While simultaneously playing the roles of croupier and data taker, he looks like an athlete, maybe a basketball player, practicing lay-ups and free throws, alone on the court late at night, stretching and jumping for the boards over and over again. After years of practice he is loose-jointed and easy, but always in control and faultless in timing each move. Spin, launch, click, buzz. We record trial after trial while accumulating data points that rise in columns high over the x axis on our histogram. If one tower in the center of the graph rises higher than the others, then we have a clear-cut advantage over the casino; we should lace the computer into our shoes and head straight for Las Vegas.

"Give it a try," Doyne says, backing away from the wheel. "You play, and I'll keep the histogram."

Stepping up to the computer, it takes me a while to get the hang of it. The data switch — a sliver of steel sprung over a contact point — is light to the touch. Trying to depress it exactly as the ball passes in front of the reference point, I sometimes click the switch too early. Other times I click too late.

"Don't worry about clicking early or late," Doyne tells me. "Just be consistent. At some point we'll hook you up to the biofeedback machine so you can work on your hand-eye coordination."

Reading the solenoids also requires some skill. The little mechanical thumpers are held in place under a layer of rip-stop nylon, and it takes me a while to gauge the difference between slow, moderate, and fast buzzes. Rather than slipping them under my belt, I find it easier to read the buzzers with my palm. Built into magic shoes, these little ticklers are eventually going to be popping off on the soles of my feet.

After a few false starts, I begin to get the rhythm. I learn not to hurry the clicks. I breathe easily and pace myself. I get better at deciphering solenoid buzzes and translating them into numbers. Standing under a bare light bulb, Doyne and I lean over the wheel for another hour. We listen to the ball rattle around the track and clatter down into the rotor's metal pockets. We're mesmerized, locked into the game so deeply that someone walking in on us might have thought he'd stumbled on the local numbers racket. This is my maiden voyage at the controls of the computer. I'm transported by the precision of the machine, the rhythm of the game, the whirl of numbers at the bottom of the varnished bowl, and the shine of the ball as it sails time and again for its rendezvous with destiny.

Rather than merely predicting the ball's trajectory, it feels as if I'm actually *steering* it. I can nudge the controls of the computer and bring the ball in for a perfect landing time and again on the surface of the spinning disk. I work the switches, and the ball drops from orbit to touch down dead-center in the predicted octant of numbers coming around to meet it. It's as if I'm *controlling* the outcome of the game, nursing the little white ball by remote control through the Newtonian cosmos of a roulette wheel whose laws I've mastered.

"Take a look at this," Doyne says, showing me the histogram for the night's play. A single tower of data points rises high over the x axis. The two columns flanking it to either side are significantly smaller. This is exactly what we're looking for — a frequency distribution in the shape of an inverted V. Most of the data points are right on target: at the congruence of prediction and outcome, the jackpot, the winner-take-all—break-the-bank beauty spot.

"According to the data," Doyne announces, "the computer is killing them. We're mopping up the house."

I imagine chips piling up in front of us. Stacks and mounds of them tossed around like Necco wafers. In this scenario the computer feels like an antigravity machine, a vacuum tube for sucking money from one side of the roulette table to the other. I grab the histogram for a closer look, while Doyne stuffs the solenoid plate under his shirt and launches the ball up on the track. There is a dangerous grin on his face as he steers the computer into another prediction.

14

Rebel Science

After Goliath's defeat, giants ceased to command respect.

Freeman Dyson

Several months after our roulette session in the Shop, I get another phone call from Doyne. "We finished the first pair of magic shoes today. I'm wearing them right now."

"How do they fit?"

"I was afraid to put them on. We still don't know whether walking on the sandwich is going to work. But I'm wiggling the switches in my toes, and the shoes feel fine."

By the time the Eudaemonic computer was ready to be compressed into a sandwich, I was no longer living in Santa Cruz. "How's the weather out there?" I ask as a newly installed New Yorker inquiring after his old hometown.

"We've been working around the clock, really jamming," Doyne says. "It looks like we'll be leaving soon for a trip into the desert. What do you think about coming out for a visit, as soon as possible?" Telephone conversations about the Project are always oblique, and this one, beamed from coast to coast, is particularly understated.

It takes me a couple of days to disappear from the city and translate myself — under the guise of a sun worshipper going on holiday — from New York to California. Walking into 707 Riverside at dinnertime, I expect to find everyone sitting down to one of their copious meals. But to my surprise the kitchen is a free-for-all of people snacking on smoothies and tortillas, and the house is in tumult.

Mobilized in a community alert, everyone is preparing for the final assault on the roulette tables of Nevada. Eudaemons walk around the house with computers on their feet. A roulette layout covers the dining room table. The click of roulette balls filters up from the basement. Everyday discourse has disappeared into technical argot about battery boats, mode maps, and histograms.

From April through the summer and into the fall of 1980 the Project had fought its way through a thicket of technical problems. There was dissension about who, if anyone, was to blame for the delays. Doyne had got sidetracked finishing his dissertation and then taken his first vacation in years — a six-week trip with Letty through Indonesia and Bali. Struggling in his spare time with the difficulties of intershoe communication, Norman had been preoccupied with the Chaos Cabal and strange attractors. Mark once again had gone out on strike, demanding more money and a larger slice of Pie.

Mark was also upset by the fact that Doyne had accepted a postdoctoral fellowship at the Los Alamos National Laboratory in New Mexico. The birthplace of Little Boy and Fat Man — the atomic bombs dropped on Hiroshima and Nagasaki — LANL has either produced or engineered most of their lethal offspring. Even though Doyne's postdoc was at the Center for Nonlinear Studies, way off in the rarefied and supposedly harmless realms of the Theoretical Division, Mark still didn't trust the arrangement. It merely confirmed his worst fears about physicists always ending up, one way or another, building bombs. He was even more upset when Norman won a NATO fellowship to spend a year studying at the Institut des Hautes Études outside Paris.

Mark demanded that Doyne phone Las Alamos and postpone his arrival for three months. Doyne agreed, realizing as everyone else did that it was now or never for finishing the Project. After their years spent perfecting the recipe, it was time to get the Eudaemonic Pie in the oven. The impending breakup of the household gave their work a special urgency. As long as the community held together, their dream of self-sufficiency remained alive. They still at the last minute might beat the wheel in a big way. This would give them the money they needed to break free of universities and governments, buy land together in the Coast Range, convert their scientific knowledge into tools for convivial living, and brainstorm a dozen other Eudaemonic projects.

The only thing standing between them and the dissolution of their enterprise was one small object: a microcomputer built into a shoe. On this slender piece of fiber glass and silicon rested their hopes for a last-minute reprieve. "I no longer have any romantic interest in playing roulette," Norman said. "I've spent enough time in casinos. But if we could make twenty thousand dollars a month, then it would be worth it. I could kiss the NATO postdoc good-bye."

On descending into the Shop the morning after my arrival in California, I find it packed with equipment and no fewer than five people seated at workbenches assembling microcomputers and building them into shoes. Norman, trying to ferret out a loose connection, manipulates the probes of an oscilloscope inside one of the radio receivers. Sine waves flash across the screen as he makes contact.

Doyne wields a solder gun over a PC board onto which he is mounting a collection of RAMs and ROMs. Sitting in a cloud of resinous smoke, he squints down at the chips and their miniature pins. "These were working fine yesterday when I had them breadboarded," he mutters, "but today they're not showing any signs of life at all."

Letty stands against the wall under the buffalo head of Manifest Destiny. Holding a piece of sandpaper, she leans over the roulette wheel and listens to the *thunk* of the swollen rotor rubbing against the stator. On the opposite side of the room, Mark faces a power saw mounted into a wooden frame. "Hold your ears," he yells. "This is going to hurt for a minute." Wearing goggles, with his hair and beard full of fiber glass dust, he makes a ferocious racket trimming the edges of a PC board. Rob Lentz, newly installed in the fall of 1981 as a full-fledged Projector, sits in the middle of the room at a workbench covered by batteries, antenna wire, connectors, resistors, and plastic boxes shaped like the heel of a shoe.

Wearing red bib overalls and no shirt over his large frame, Lentz is a blue-eyed southern Californian in strapping good health. He sports a mustache and medium-length brown hair parted in the middle and puffed around his ears. With strong arms and a layer of adipose evenly distributed over his body, he has the well-scrubbed look of a surfer newly beached.

In spite of ten years' difference in age, Lentz and Truitt were in the same class together at UC Santa Cruz. Doyne had been Lentz's teaching assistant for a course in mathematical physics, and Rob

Shaw had been his adviser for a senior honors thesis. When Mark went for his job interview at Watkins-Johnson, the classmate whom he met there was Rob Lentz. And when Mark walked out in disgust on learning that it was a military assignment, it was Lentz who had landed the job. Rob stayed at Watkins-Johnson for a year and a half, working his way up through the ranks to become project manager on a $21 million contract to Raytheon for Sparrow missile components. Then *he* walked out in disgust.

"When I told them I was quitting, the big bosses took me to lunch and pleaded with me. 'Rob, we don't understand. This is our plum contract. Do you want more money? Do you want more people?' They were falling over themselves to find out why I was leaving. I worked in the VCO division, which was the fastest-growing part of the company. VCO stands for voltage control oscillators. These are high-frequency, small-wavelength radar devices — the same sort of stuff we're making down here for the Project — but at Watkins-Johnson everything was done half-assed. When they put me in charge of the missile program, it was already two years behind schedule. We were filling orders for military hardware, which meant we had to spend as much time trying to break the stuff as make it. This was your run-of-the-mill, medium-sized electronics company. But the place was a mess, and if that's the way the electronics business works, then I don't want to be part of it.

"At my lunch with the bosses I told them straight out, 'I don't like working for the military. I don't want to argue the morality of it. But I'll let someone else do the dirty work. It's leading straight to Armageddon.' A lot of seemingly good physics ends up in the wrong hands; so I'm thinking of getting out of science and doing art. My brain has a right hemisphere that isn't getting used enough."

Rob considered going to medical school or graduate school or on a cruise to Mexico. He tinkered with a couple of patent ideas and audited Rob Shaw's lectures on chaos theory. One afternoon he was walking out of the physics building to catch some surf when he ran into Doyne in the corridor. They chatted for a few minutes and then Doyne said, "As long as you're unemployed, I have a business proposition that might interest you." They drove into town for an espresso at Caffè Domenica, and Doyne told Rob about the Project.

"The story didn't sound believable. So we walked across the river to the house. When I saw the Shop and the roulette wheel and the computers, I thought, 'This is really neat. It's the same stuff I've been doing at Watkins-Johnson, but it's anti-nuke physics. It's high tech for civilians.' I was looking for something like this and it arrived at just the right moment — a chance to do rebel science."

Rob signed up to work on the Project full time for no pay and nothing more certain by way of remuneration than a slice of Eudaemonic Pie. He had enough money for the moment. He could knock off in the afternoons and go surfing. All he cared about was putting his knowledge to use doing something other than building Sparrow missile components. His good-natured appearance down in the Shop came like a shot in the arm for the Project. Mission Control had unexpectedly sent up another good man to join the team already in orbit.

A year and a half had passed since Doyne and Norman first imagined building a computer in a shoe. Once Mark had joined the Project, the three of them had worked steadily on the new device, which was now finally ready for installation. They had come up with a three-footed system in which the data taker wore a mode switch and low-frequency radio transmitter under the arch of his left foot. In his right shoe he wore a data switch, a microprocessor built into a computer sandwich, three solenoids, a collection of batteries, and a transmitter capable of beaming predictions to a bettor standing up to ten feet away. The bettor's right shoe — the third foot in the system — contained a microprocessor computer sandwich, a radio receiver and antenna, a battery unit, and three solenoids designed to vibrate, like those of the data taker, under the arch and heel of the foot.

To pack them into such limited space, the transmitters and other components had been built into special containers shaped like the soles and heels of shoes. Designed for slipping in and out of cavities cut into false-bottomed footwear, these modular units simplified the process of recharging batteries, troubleshooting the system, and otherwise adjusting the computers for a comfortable fit.

On finishing his silicon sculptures, Mark looked at this new form of ambulatory art and declared, "The components, especially the computer itself, are so solid and so appropriate to their purpose that I find them aesthetically pleasing. They have integrity as objects. I look at them and say, 'Is there any better way to do this?' and I see there isn't."

Once Mark had built a prototype version of the system — complete with mode switch, computer sandwich, and battery unit — all that remained was the topological chore of designing a shoe roomy enough to carry them. On returning from his trip to Bali, Doyne asked around town for a shoemaker who specialized in custom work. He found one down the coast in Rio Del Mar, and phoned for an appointment. The Project was a closely guarded secret, especially at this delicate stage of the operation. So Doyne carried to his meeting with the shoemaker not the computer components themselves, but blocks of wood cut to scale. He planned to say nothing about their use.

On driving to Rio Del Mar, Doyne was surprised to meet a man in his early thirties — tall, tanned, and evidently as hip as many of the other young artisans in the Monterey Bay area. "I need a pair of false-bottomed shoes," Doyne said. "Nothing illegal, you understand."

The man stepped to the front of the store, locked the door, and pulled down the blinds. He led Doyne into a back room and asked him what he had in mind. Doyne produced his blocks of wood and said, "I don't want to go into the details, but I need to walk around with these fit into a pair of shoes. This big piece goes up front under the toes. This other piece rests behind it under the heel. If you can do that, then you should have no problem getting this smaller unit into the left shoe."

"First of all," said the shoemaker, "I don't care what line of work you're in. Drugs, gems, it's all the same to me. If I don't ask any questions, you don't give me any answers. That way you keep your business to yourself, and I keep my nose clean.

"Take a look at this," he said, pulling a pair of walking shoes off the shelf. "They look like your everyday strutters, don't they? There's nothing special to them, until you peel back the insole. That's right," he said, as Doyne looked inside. "There's a lot of cargo room in there.

"It requires some gluing and stitching, but a job like this is relatively straightforward. All you need to do is go out and buy the right kind of shoe — something I can take apart and put back together again without any sign of its being tampered with. You could order custom work and build a shoe from the ground up, but that'll cost you. That's why I recommend going with the ready-mades."

They shook hands on the deal and agreed to meet the following day in Santa Cruz. In exchange for a guided tour of the local shoe stores, the cobbler wanted Doyne to buy him lunch at Hilary's, the most expensive restaurant in town. "I'm sure he had no idea what we were doing," Doyne said. "He probably thought we were transporting drugs, because he told me once he wouldn't mind receiving a little 'present' on our getting back from our trip.

"This guy turned out to be quite a figure in the community. He showed up for lunch wearing an ascot and sports coat. The waitress must have practiced for weeks to get her smile so perfect. A steady stream of people kept coming by the table to say, 'Hi. How ya doing? I'd like to drop by the store tomorrow and show you a little something.'"

After lunch, the shoemaker, Rob Lentz, and Doyne walked down the Pacific Garden Mall. They paused in front of a dozen window displays, while the cobbler offered a running commentary on the nature of footwear. "Penney's specializes in ladies' tassled tennies and nursing whites. Gallenkamp's, favored by the welfare set, offers your eight-ninety-eight all-plastic special. At Morris Abrams you can get your Padmores and Florsheim's with real crepe soles and price tags over sixty dollars. These other stores go in for spaghetti-Western cowboy boots, high-heeled moccasins, desert kickers, Birkenstock sandals, and Gucci loafers. But you'll also notice, if you look around, that a lot of people have said, 'To hell with the whole thing.' So they're walking down the Mall barefoot."

Stopping at Herold's shoe store, which had a window display of loan-officer six-eyelet brogans and Pat Boone loafers, they also saw some Dex and Drifter crepe-soled shoes with the flared toes made popular by the healthy-foot movement of the 1960s. Several jauntier models, influenced by Nike running shoes, looked like duck feet with racing stripes on them. On walking inside, the shoemaker picked a dozen shoes off the racks and sat surrounded by them in the middle of the store.

"The first thing you need is depth in the sole. So you want to go with crepe or one of the better synthetics. Watch out for this kind of waffling," he demonstrated to Doyne and Rob, as he flexed the bottom of a shoe with air holes layered into it. "The sole isn't solid, and when you cut it, it just falls to pieces. You can buy a shoe with a heel or a straight wedge, but in either case you have to have a slipsole. That's this thin layer inserted between the sole and the upper. Without it, there's no way you can stitch the shoe back together again.

"Most of the stuff in here is junk," he said, peeling back the tongues of shoes and staring inside. "Instead of making shoes with slipsoles and stitching, they just fold the leather over and glue it down. The rest of the story you can pretty well figure out for yourselves," he concluded, bending a shoe in half. "You want something flexible but solid, with plenty of toe room and no metal shank or other obstructions."

A saleslady hovering nearby alternately frowned and asked, "Can I help you?" She looked nervously at the three of them. Beside the shoemaker, dressed in his sports coat and ascot, Doyne was wearing a Balinese tie-dyed T-shirt, shorts, and a pair of down-at-the-heel jogging shoes. Rob, the third member of the party, sported buffalo-chip sandals and red bib overalls, which made him look like a farmboy hybridized out of corn pone and cannabis.

"We're in town making a movie," Doyne told the woman. "And we have some scenes that require shoes with special effects."

They bought a pair of Bass walkers and another pair made by Clarks. Three days later the shoes reappeared at Riverside Street, apparently unaltered except for the substitution of wedge soles. The shoes from the outside looked meticulously normal. It was only on lifting up the insoles and peering inside that one discovered cavities large enough for a major drug run from Colombia — or the insertion of mode switches, battery boats, and computer sandwiches.

The week before Halloween, the Project picked up speed. Letty quit her job and moved back to Santa Cruz. Doyne phoned his boss at Los Alamos and broke the news that he was going to be three months late for work. Rob Lentz gave up surfing in the afternoons. Mark Truitt spent sleepless nights fine-tuning the system. Norman shelved his dissertation and went back to ferreting bugs out of the radio receivers. Rob Shaw played background music on the piano. Grazia Peduzzi cooked pasta for everyone. Lorna paid the bills and tended the garden. Wendy Tanizaki worked three jobs at once so that Mark could stop worrying about money. Everyone started talking the language of computers. Conversations were brief, tactical. At Project meetings held around the dining room table, last-minute tasks were doled out three and four at a time: load chips onto PC boards, build battery boats, tune up receivers, oil the wheel, visit the shoemaker, shop for chips in the Valley, practice on the eye-toe coordination machine, design wardrobes for Las Vegas. Like a theater backstage before opening night, the air thickened with tension and nervous good humor.

To get computers and players to Las Vegas as soon as possible, the Projectors considered a two-wave approach, with one team leaving for Las Vegas immediately and the second following when more equipment was finished. Doyne was the obvious choice to fill the data taker's shoes and go off in the first wave. But I was surprised to find myself nominated and elected as the second member of the team, playing the role of high-stakes bettor.

"We'll dress you in a cowboy shirt and string tie," Doyne said. "You'll do great with a southern drawl. You look like there's money in your family."

"But watch out at someplace like Caesars Palace," Letty advised. "When they turn on the heat, it can get hot fast." As soon as you start winning at roulette, a lot of attention comes your way. Pit bosses and floormen nudge in at your elbow. They go from solicitous to threatening as the temperature mounts. At this point in the scenario it becomes a bravado performance, a high-wire act over an abyss of potentially very bad news. But the skills required for playing roulette in Las Vegas with a computer on your foot — stamina, wit, duplicity, and the reflex of technique perfected to invisibility — aren't these the same skills that a writer summons up on facing the blank page? This fond thought, as delusive as it may have been, reminded me that I needed to invent a story about myself, one that Las Vegas croupiers would be reading over my shoulder with great interest. For such a discriminating audience, I had to produce a narrative that flowed without a stutter from beginning to end.

"O.K.," I reply, addressing my fellow Eudaemons officially for the first time. "Give me a pair of magic shoes and I'll play the role of high-stakes bettor. But we'll have to rethink the part. I've never liked Texas string ties. How about substituting a gold chain and a pinky ring?"

Before deciding definitively on the two-wave approach, we schedule another Project meeting for the following night. After dinner, Doyne, Letty, Norman, Rob Lentz, and I walk down the back stairs, across the garden, and through the high wooden gate at the far end of the yard. We come out by the barn belonging to the Riverside house, where a common driveway serves a collection of bungalows built under the levees along the San Lorenzo River. Down the driveway and around the corner we come to Mark and Wendy's cottage.

With hardwood floors and white walls, their two-room house is spare and ordered. The front room holds a dresser, a small table, a potted fern, a quilt-covered bed, and a Goines lithograph of "Pandora's Box." A bookcase along the wall is filled with textbooks ranging from biology to optics, an *Encyclopaedia Britannica*, a collection of science fiction, an oil lamp, and a jar full of paint brushes. The six of us squeeze into the room by sitting on the bed and floor. Holding a clipboard of notes written on a yellow legal pad, Doyne takes up the agenda.

"There's a long list of things we have to do before we can get out of town. I think we should divide up the tasks and put our initials next to them. That way we'll know who's responsible for doing what." Mark says he needs three days to finish the first two computer sandwiches. Rob reports thirty to forty hours' work remaining on the battery boats. Letty claims she is making good progress on assembling the solenoids. Norman is optimistic about getting the radio receivers debugged momentarily. Doyne signs up for finishing the mode transmitter. I take responsibility for teaching myself the layout and betting patterns. At the end of the discussion, Doyne tallies up the list. "There are a lot of hours here," he reports. "Probably on the order of five or six more days."

"Watch out for Murphy's law," Rob advises. "If anything can go wrong, it probably will."

"We have another forty or fifty hours of practicing left to do," Doyne says. "I want to set up the eye-toe feedback device and gather histograms. I also want to figure out analytically what advantage the new computers have over roulette wheels that are going really fast, in case that's what we run up against in Las Vegas."

"Why can't you figure out our advantage from histograms gathered during the practice sessions?" Letty asks.

"Because in physics, it's best to come at questions like this from both directions. You want to figure out the answer empirically *and* theoretically."

There follows a long discussion on how many computers the first team should take to Las Vegas. Doyne wants two computers for a complete system, and another two computers for a back-up. Mark is in a hurry to send the team off with only one set. "I'm queasy about going to Nevada without a back-up system," Doyne tells him. "Up on the bench the computer works fine. But when I start walking around on it, it tends to flake out. I get ghost signals and other garbage in the clicks."

"One also wonders," Norman muses, "whether there is something about the electronic environment in Las Vegas that might make the computer unhappy. Where do you think the system is vulnerable?" he asks Mark.

"We don't have any software protection against noise," he answers. "There could be a lot of garbage in the environment that confuses either the computer or the radio transmission."

"In the past we've always gotten squeezed in Las Vegas," Doyne says. "Even when we went with two or three sets of equipment, we invariably got pared down to only one that worked. So people sat around getting antsy, and the shift rotation turned into a mess."

The question of how many computers to send with the first team remains open as Doyne turns to other items on the agenda. "Someone has to figure out how much money we should take for betting capital. We need a repair kit. We have to make costumes. And then there's the question of transportation. Letty's Fiat is a ringer, but finding a second car is a problem. What do you think about taking the Blue Bus?"

"That's a bad idea," Rob answers. "With an orange sticker on the front saying, 'Question Authority,' the Bus doesn't really project the kind of image we want to have when driving up to Caesars Palace."

It is past midnight when Doyne reaches the end of his list. "We need to call a dress rehearsal as soon as possible," he concludes.

"I'll be the pit boss," says Mark. Norman offers his services as croupier. Doyne volunteers as data taker. Rob says he'll fill in as a cocktail waitress. "I'll be the cashier," says Letty. "And Thomas," she adds, turning to me with a smile, "you get to be the roulette player stealing money out from under our noses."

I walk into the Shop early the next morning to find everyone already at work breadboarding computer sandwiches, assembling battery boats, and spinning the wheel. In a room no bigger than twenty feet by twelve, with walls either unfinished or covered merely by the silver lining of Johns Manville insulation, every available inch of concrete floor is taken up by the two posts needed for holding up the roof beams, three workbenches, a drill press, a grinder, and the roulette wheel jacked up on its table. New to the decor are a wet suit hanging in the rafters and a surfboard leaned against the wall under the glowering head of Manifest Destiny.

"That's for taking the system to Biarritz," says Rob, pointing to the surfboard. "When we're not playing roulette, we can catch a few waves."

I remark that the room looks neater than usual. The plastic ice cream containers lining the shelves have been newly labeled MSC CHIPS, LEDs, MSC RESISTORS, TRANSITORS, 110 V AC 60 Hz MALE AND FEMALE PLUGS, and so forth.

"It's Rob's good influence," Doyne says. "He even makes us put away our tools at the end of the day."

Doyne sits at a workbench covered by a Tektronix oscilloscope, a signal generator used for tuning components, and a power converter needed for transforming AC to DC and stepping it down to the meager 5 or 10 volts on which computers run. Lying in the middle of these tools is a breadboarded mode switch.

"Yesterday it worked great," he says, "but today I can't get a clean signal out of it." He probes among the components with needles wired to the oscilloscope. Instead of uniform waves rolling over the face of the scope, the lines are roughed into peaks and troughs. Hoping to straighten them out, Doyne spins a couple of dials under the screen, but this only skews the waves into gale-force chop.

"Batten down the hatches," Rob yells. Wearing green safety goggles, he leans over the drill press and drills three neat holes along the edge of a battery boat. Designed to fit into the heel of a shoe as the power supply and radio unit for a computer sandwich, each boat (so named because its rounded shape makes it look something like an inland scow) holds several kinds of batteries, an antenna, two solenoids, and a radio transmitter and receiver.

"This is a little tricky," Rob comments after drilling the holes. "There are a hundred and forty turns of antenna wire embedded in these things. One false move with the drill, and you end up tossing a week's work in the garbage bin."

The battery boats not only resemble boats, they are also manufactured like them. The process begins with a plaster mold in which antenna wire, looped and tied together, is strung along the outer edge. After layers of spun-glass cloth have been laid over the wire to act as reinforcement, the mold is filled with liquid casting resin bought from a marine supply store. In half an hour the resin has hardened to the consistency of gelatin, and by the end of a day it can be walked on by an elephant, or at least a human.

When broken out of their plaster molds, the units are ground, routed, polished, drilled, and loaded with components. The boats end up carrying a full cargo of resistors, capacitors, diodes, batteries, solenoids, antennas — in short, everything needed to power a computer, as well as to input, output, and transmit its signals. After being tested, the boats are fitted with a clear plastic lid, out of which stick two of the system's three thumping solenoids. In order to register on a different part of the foot, the third thumper lies farther forward on the sandwich.

"This isn't your everyday plastic," Rob explains, holding up one of the boat lids. "It's polycarbonate, or Lexan. At nine dollars a square foot, it's the same stuff they use to make jailhouse windows."

Complete with a ribbon cable trailing out its stern, a finished boat looks like a teaching aid in a high school biology class — maybe a sperm magnified a million times. The ribbon cable, which is designed to stretch along the arch of the shoe, connects the boat to its computer. Attached to the end of the cable is an eight-pronged socket with individual pins capable of delivering either 5 or 0 volts (required for running the microprocessor) or 20 volts (needed for radio transmission and solenoid buzzes).

"These sockets are actually made for flying model airplanes," Rob says. "They're nifty little things. But getting the model airplane connectors soldered onto the cable has to be the biggest pain in boat building."

Holding a grease can and a collar of ball bearings lifted from the internal mechanism, Letty leans over the roulette wheel. "We were getting lousy predictions yesterday," she says. "The rotor is swollen from having been rained on this winter. So I'm taking it apart to sand the rim and grease the bearings."

"It's too humid for it in Santa Cruz," Doyne remarks. "The wheel would be happier if we took it back to Las Vegas."

"After all the things that have been done to it," says Rob, "I wonder if the wheel still likes us."

"It's had a more exciting life than most wheels," Letty says. "We liberated it from the casinos and brought it out here to do something much more interesting. I'm sure the wheel still likes us."

With the radio tuned to KFAT, Dolly Parton comes on the air singing the theme song from *Nine to Five*. Mark Truitt, his hair and beard tufted into clumps, strides into the Shop. "Waxed and ready to go," he exclaims. "After a hike to the Mall and back, the sandwich still scopes out perfectly."

Mark holds in his hand a fiber glass rectangle tapered at one end. No larger and barely thicker than the sole of a size four shoe, the computer is opalescent gray and slightly transparent, so that on holding it up to the light one discerns inside the handful of smaller black rectangles that have been layered on top of each other in this electronic BLT constructed out of RAMs, ROMs, and a microprocessor.

The problem with Mark's metaphor, as just demonstrated, lies in the fact that this particular sandwich is sturdy enough to walk on. After inventing a novel design for building computers with their chips inverted on top of each other, Mark has come up with another original idea for fixing these components in place by means of a process he describes as waxing the sandwich. Given the idea before he discovered the means to implement it, he had imagined that somewhere in the world there existed a substance, viscous when heated, hard when cooled, that could be poured into the middle of a computer sandwich. This miraculous material, after flowing around the microprocessor, would cool and set up like concrete with a time capsule buried in the middle of it. The only parts of the computer still exposed to the outside world would be a toe-operated microswitch, a solenoid, a battery plug, and the backsides of two PC boards, Acting as top and bottom to the sandwich, the boards would be covered with the solder points of an otherwise inscrutable circuit.

Working in favor of Mark's idea for waxing the sandwich was its simplicity. It made for a one-piece, ready-to-wear computer. Should it ever fall into the wrong hands in Las Vegas, the device would likely remain an unidentified object that not even a specialist could reverse engineer. But the fact of its impregnability also argued against Mark's idea to wax the computer. Once sealed, troubleshooting the system and replacing burnt-out components would prove difficult, if not impossible.

The question was argued back and forth at several Project meetings. Doyne favored leaving the sandwich open and building a metal box around it, while Mark wanted to seal the computer in perpetuity with a filling of epoxy resin. They compromised on a third solution. After fitting a polycarbonate spacer between the upper and lower PC boards, the sandwich would be filled with microcrystalline wax. This was the magic substance Mark had been looking for. A petroleum product related to plastic, microcrystalline wax is hard, strong, and brittle — except at 300° Fahrenheit, when it takes on the consistency of molasses.

"It's a daring thing to do," Mark says, handing me the newly waxed computer. "You take a microprocessor that works, and in an hour you make it very hard to fix if you've messed it up."

Explaining the waxing process, Mark describes how he stayed awake most of the night trying to calibrate his gas oven to exactly 300°. "That's the maximum rated storage temperature for our chips. So I didn't want to take any chances. I had a thermometer inside the oven, but it was tricky keeping the door wedged open to read it. When I had shut all the air currents out of the kitchen and finally got the temperature just right, I popped the computer into the oven for an hour. I wanted to warm it up before pouring in the wax, which I was melting on the stove in a double boiler with a candy thermometer. When the wax was highly viscous, I dripped it into the sandwich through a funnel and let everything bake in the oven for a few more minutes. Then I took out the computer and laid it on a cooling rack. As soon as I could, I put the scope to it and prayed everything was still intact. If there had been any problems, I theoretically could have thrown the computer back in the oven and melted the wax out of it. The process is supposed to be reversible, but we haven't had a chance to test it yet."

Holding the sandwich in my hand, the only parts of the computer still clearly visible are the copper tracings etched on the inside of its PC boards and the nodules of solder that dot the backsides of these boards. The solder points indicate where under the fiber glass lie the golden pins that lead to buried boxes of silicon, boxes now frozen possibly forever into a translucent sea of microcrystalline wax.

Mark points to the curving lines inside the sandwich. "This is unusual architecture for the circuit of a computer. You'll notice that none of the lines is straight."

"Why is that?" I ask.

"Because I drew them without a ruler."

15

"Dear Eudaemons"

The zodiac may be regarded as an immense roulette wheel on which the Creator has thrown a very great number of small balls.

Henri Poincaré

That night — actually, in the small hours of the following morning — I find Doyne and Norman standing at the kitchen table eating gingersnaps. Scattered in front of them are pages of graph paper covered with hundreds of data points.

"I'm nervous about these," Doyne says, pointing to the graphs. "Letty worked all day collecting data. But it seems more random than usual. I think I'll pull out the old histograms and see how they compare. I also want to run the KIM in place of the shoes. By using the computer to simulate our toe clicks, we should be able to find out whether the statistical fluctuations are due to human error. But what makes me suspicious is the fact that Joe Random — some space case from down on the Mall — should be able to click better than this."

Norman chews a gingersnap and strokes his beard. "Has anyone tampered with the program?" he asks.

"Mark made some changes so that now we clock the rotor past the reference point at every other revolution. It's more accurate that way. And he's done some housekeeping at one or two other addresses."

"Does Mark have any theories about what's going on? Is he thinking about the problem?"

"Yes, he's thinking about it. But at this point, I'd rather not have any theories. We should run more tests and put the program on the KIM."

"At least you have some help in the software wars. You used to be all alone out there."

"I wish I still were," Doyne says. "There's no documentation for any of the recent changes in the program, which means there are stretches of it that *no one* knows anything about."

Later that morning, after breakfast, Doyne clears the dining room table and turns to me. "We're picking up your shoes from the cobbler this afternoon. You'll have to use your fingers until then, but I think it's time you got started on betting practice."

He spreads in front of me the Eudaemonic layout. Out of an old cigar box he pulls a handful of plastic casino chips and scatters them across the baize. "It may not be the real thing," he says, "but you get the idea."

Next he places on the table a computer sandwich and battery boat. The two units, nestled one behind the other, look like the sole and heel of a shoe designed for a clubfooted toddler. Walking on them will take some getting used to. Doyne stretches the ribbon cable from the boat to the computer. "As soon as you plug in the sandwich, you're powered up. The batteries should last several hours. But be sure to unplug the computer when you're not playing."

Beside the cigar box full of roulette chips Doyne places a second, larger container made out of green plastic. "This is the betting practice box," he says, propping open the lid. "Inside you have an LED display, some batteries, and a few chips wired into an electronic circuit."

I peer inside the box. "That's Harry in there," Doyne announces, referring to the old roulette computer. "I was sorry to see this happen, but Mark thought we should cannibalize it."

Doyne explains to me how the box works. "When you flick this toggle switch, Harry starts transmitting signals strong enough to buzz your solenoids from fifteen feet away. I designed a special random number program for the circuit, so there's no pattern to the buzzes."

A tinking noise comes from the computer and battery boat in front of me. One of their three solenoids begins to pop up and down. Doyne places his hand over the buzzes and dampens the sound against his palm. With one solenoid sticking up from the sandwich and two from the boat, the buzzers are lined up within a few centimeters of each other. I hear another *tink, tink, tink* come from under his hand.

"That's a one," he says. "A low-frequency buzz on the front solenoid." I look at the LED display inside the betting practice box to see a red number "1" light up in computer script. A second, more insistent *bzzzz* comes from under Doyne's palm. "High frequency, front solenoid," he says. The diodes inside the box light up with the number "3." Another *bzzzz* comes from down near Doyne's metacarpals. "High frequency, middle solenoid." A red number "6" glows on the display.

"It's like a computerized flash card," he explains. "I designed the box for training bettors while driving to Las Vegas. The speed is variable. By toggling this switch, you can make the buzzes come at you as fast as you want. After you've learned the signals and picked up speed, then you can start to work on your betting patterns.

"You have to memorize the numbers in each octant on the wheel, as well as one or two numbers to either side. This allows you to vary your betting patterns and confuse anyone trying to figure out your system. You want to cover only three or four numbers at a time, depending on your bank-to-bet ratio, but you need a range of choices. If the table is crowded and you're stuck at one end of the layout, you might have trouble reaching all the numbers. If you can't get to 30, for example, you can substitute 11, which is next to it on the wheel, but farther up the layout. You have to work out various strategies like that for getting your bets placed as fast as possible. Your moves should be automatic. The computer gives you a moderate buzz on the middle solenoid. Octant five. Numbers 12, 8, 19, 31, and 18. Wham. You cover 8 and 12, because they're next to each other on the layout, and then you hit 18 and 19. But you might want to forget 31, because it's way down at the far end of the table."

Doyne leaves me in the dining room standing over a tablecloth painted with red and black numbers. I place my palm over the computer and battery boat. The front solenoid tickles my life line as it pops up and down. *Tink, tink, tink.* A low-frequency buzz. Octant number one. I look into the betting practice box and wait for the diode display to light up with a squared-off computer digit. I smile to myself on seeing the electronic flash card give me a red thumbs up.

I lock into playing roulette and barely notice throughout the day as one storm after another rolls in off the Pacific. Late in the afternoon, I walk into the garden to find strong winds and thunderheads building up into another storm. Entering the Shop, I discover Letty, Rob, and Doyne huddled around the roulette wheel. This mystic icon has every inch of its varnished frame covered with wires and scientific instruments, including a handful of optrons pointing down onto the ball. These are infrared, photosensitive transistors — neat little gadgets that record the position and velocity of a moving object by means of infrared radiation. I watch Letty clicking microswitches and then turn to look at the red numbers glowing on the face of the KIM.

"We're running Letty on the hand-eye coordination program," Doyne says. "This is the old Human vs. Machine experiment, where you get to see how spastic you are in clocking the ball compared to an optron. Letty is basically doing fine, although there's something funny going on, and we can't put our finger on it. We set the parameters and the predictions look great. But then half an hour later we find the ball falling short by an octant or two. For some reason the predictions are drifting."

Throughout the past week, heavy weather has been blowing in from the west. The sun shines and then disappears in a chop of clouds and wind across the Bay. Waves roll over the town pier and wash up on the boardwalk, while out in Steamer Lane only the best of them are surfing the swells. The weather has been fickle under a full moon, undecided between clearing into blue skies or sending more clouds scudding up the San Lorenzo River valley, where they hang in the redwoods like a soggy tent. With a thunderclap and torrents of rain, another storm breaks outside.

Mark pushes open the Shop door. Wearing only a T-shirt and pants, with rainwater dripping off his beard, he shakes himself like a dog. "I've got it," he exclaims. "I found the problem. Take a look at this." He picks up a half dozen roulette balls and stands over the wheel. In rapid succession he spins all the balls at once. They differ in size and shape and in the sound they make circling the track — from the taut, high-pitched note of plastic, to the mellower sound of ivory on wood. They also vary greatly in speed. The fast ones bump into the slow ones, bounce backward, and catch up again to the laggards.

"At the widest spread," says Mark, "the Teflon ball decelerates a hundred percent faster than the composite ball. That gives you an idea how sensitive these things are. We're setting the parameters all right for the different balls, but the balls themselves aren't staying the same. The problem is the air. I mean air isn't just air. Some of it's more viscous and harder to travel through than other air. I notice this when I ride my bicycle on a foggy day. I was watching Rob and Letty making histograms this morning. Everything was working great, until a half hour later they found their predictions slipping off center. The longer they played, the farther the ball traveled away from its original exit point. I went home to think about it and came back during lunch to conduct a little experiment. It was sunny outside, and I left the Shop door open. The parameters for the ball began to drift. I shut the door. They swung back an octant. The ball was clearly traveling farther with the door open.

"I went home again and phoned Bill Burke at the university. I knew he played billiards and thought he might have some advice on what was happening with our balls. He told me this kind of problem is familiar to anyone who's played billiards with the original ivory balls. They actually change shape and go out of round with variations in barometric pressure; tournaments used to be delayed until the pressure stabilized. Since we're dealing with acetates and plastics our problem is a little different, but he still thinks that what we're noticing is caused by alterations in the pressure and viscosity of the air. I figure we're getting a five to ten percent drift every hour. We always knew that different balls had different rates of deceleration, but no one guessed that the *air* could change so fast in so little time."

"It should be better out in the desert," Letty speculates. "I imagine the casinos are climatically more stable."

"Mario Puzo tells a story in *Fools Die* about a casino owner wanting to liven up the action," Doyne says. "So at three o'clock every morning he'd pump in a few tanks of pure oxygen. Who knows what kind of ball drift you'd get then."

*

As the storm passes and the sun reappears, Doyne turns to me. "Let's bag some endorphins," he says. We put on our track shoes and sprint along the levee overlooking the San Lorenzo. The river is fast and muddy. Ahead of us a wall of clouds is pushed against the mountains. We cut through the town cemetery and slow for the long climb up into the hills. The air is thick with the smell of eucalyptus and bay laurel. The bark on the manzanitas shines burnt orange under a varnish of rainwater. On reaching the fog line, where patches of mist drift among the redwoods, we turn back toward the ocean.

Below us, straggled over the flood plain of the San Lorenzo, lies the town of Santa Cruz. Among the major landmarks, we make out the Mission church, rebuilt after a fire to three-quarters scale, the Ferris wheel turning over the boardwalk, and the big "D" illuminated on top of the Dream Inn. Dark on the edge of town are the Brussels sprout fields to the north and the forests and clustered campuses of the university. A touch of red neon glows in front of the fishmongers on the wharf. The lighthouse blinks over the surfers in Steamer Lane catching the last waves of the day. Directly in front of us lies the sweep of the Bay and a hook of land on the far side that sparkles at night with the lights of Monterey. As we sprint home down the levee the clouds above us break into patches of blue, while out in the Bay shafts of sunlight illuminate the golden V's of fishing boats towing their wakes back into harbor.

For dinner Grazia has cooked pasta and chicken breasts in cream. There is a big commotion tonight over Norman, who is making his debut singing in a concert of Renaissance music. Lorna bustles around getting him dressed. "Packard, it's amazing what you think is cool," she says of his first attempt. "It's not cool at all." Norman reappears wearing skinny black pants and a beige, open-necked shirt. Lorna wraps a scarf around his neck. We rush through dinner and bundle him out the door.

The church is filled to overflowing as Norman comes on stage. Looking like a choir boy sprouting through prolonged adolescence, he sets the pitch and tempo for the opening madrigal. His tenor is sweet, but not yet warmed enough to carry through the hall. A chorus and solo voices, accompanied by an orchestra of lutes, cornets, sackbuts, violas, and a small organ, perform what the program notes describe as "Music of the Serene Republic." A mixture of church and secular songs composed in Venice during the seventeenth century, the music expresses an order and coherence that have long since disappeared from the world. Grazia, sitting next to me, lapses into Italian. *"La stella,"* she exclaims over the soprano. *"Brava. Brava."* This is the music of a free people, say the program notes, who were far enough removed from the center of the Roman Empire to cultivate an aesthetic sensibility and refinement of their own. "They think they're describing Santa Cruz," says Grazia, laughing. "Far from Washington and New York, this is the Venice of the American Empire." Whatever truth there may be in the analogy, there is indeed a resonance and lightness to the music that corresponds perfectly to the graciousness of a mild night on the shores of Monterey Bay.

First thing in the morning, Doyne, Norman, Mark, Letty, Rob and I gather in the dining room for a Project meeting. Norman, yawning hugely, is wrapped in a red dressing gown. Mark, wearing a long-sleeved T-shirt, chinos, and Nike running shoes, squats on the window bench, where he bounces on his toes like a football coach hunkered down on the sidelines. Letty is in blue jeans and a cotton shirt rolled up at the sleeves. "Last night I dreamed that bubble gum got stuck on the wheel and melted down into the bearings, where it totally gummed up the works," she says. "God, it was awful." Rob Lentz shows everyone his new haircut and freshly trimmed mustache. "I'm cleaning up my act for Las Vegas," he reports.

Doyne, wearing Patagonia pants and an Icelandic wool sweater, his hair still wet from a morning shower, opens the meeting by handing out photocopies of "Predicting Roulette," the Project's twenty-five-page manual on how to beat roulette with computers. "You've seen this before," he says, "but some of you were missing a few pages. I've also thought of a neat improvement for the program, a way to adjust one of the parameters automatically while we're playing. It would take a couple of days to implement, and I don't know whether we want to wait around that long."

Nailed to the dining room door is a blackboard that Doyne covers with equations while launching into a miniseminar on the physics of roulette. Long strings of variables unfurl between the brackets of logarithmic functions. Deltas pop up in front of adjustable parameters. Parentheses fence off measurable rates and periods. Preset values and "fudge factors" are used sparingly throughout. Chalk dust floats in the sunbeams slanting in at the window as Doyne finishes writing the last of three multifunctional equations on the board. "These are the two equations of motion for the ball and the rotor," he concludes, "and here's the solution, which is an algorithm that combines these equations and solves them."

After writing down the equations in a notebook, Rob looks up and strokes his mustache. "Doyne," he asks, "do I really need to know all this stuff? I mean, so long as the hardware works, all I want to do is get in there and use it."

"Hold it," says Mark, rocking back and forth on his toes. "I'm really paranoid about this. Every time someone talks about something going wrong with the system, they blame it on the hardware. But our screw-ups have come just as often from software."

"Boy, you're jumpy today," says Rob. "I wasn't blaming you for anything. I was asking Doyne if I could play roulette without knowing all the physics. I'd *like* to know the physics, but I thought we wanted to get out of town as fast as possible."

"I agree with Rob," says Letty, who is curled up in a beanbag chair in the corner of the room. "I really haven't been following much of this. What we need right now is a benevolent dictator. Why don't you just tell us what to do, and we'll go do it?"

"I think two of you should leave for Las Vegas as soon as possible," Mark insists, "without changing the program or waiting for a second set of computer sandwiches to be waxed. I know you want a back-up system," he tells Doyne. "But I can get one finished and sent out to you in a couple of days. You should forget about the extra testing with the feedback device and get on the road."

"But it's been a couple of years since we ran a complete set of eye-toe coordination tests," says Doyne.

"I know," Mark agrees, "but how badly has your nervous system deteriorated since then?"

I am wearing a pair of magic shoes, fully loaded, as I stand in the basement getting buzzed with signals from the betting practice box. Clarks, nice looking walkers, there is nothing unusual about these shoes except for the computer inside that's tickling my right foot.

This is my final fitting — or "tweaking up," as we call it — to get my buzzers adjusted correctly. Letty has become expert at manufacturing the metal plungers that bob up and down in the solenoids. As the last step in the process, she form-fits the plungers by filing them down to points. There is a knack to wearing a computer in your Wallabees. You want to walk without a gimp, but tread gingerly on the microprocessor underfoot. A slight lift to the heel allows the plungers to jump like popcorn. But after the split second it takes to read them, a push on the tarsus can damp down the solenoids completely.

Wearing his own pair of magic shoes, Doyne paces around the roulette wheel. "My computer went bye-bye. I just got a nine," he announces, referring to the "no-bet" buzz, a high pulse on the back solenoid. "As soon as I put on the shoes they stop working."

"Do you have smelly feet?" Rob asks. "I've been thinking we should run a smell test on the computer."

Doyne's face breaks into a lopsided grin. "Yeah," he agrees. "We should test everything."

The other joke of the day has to do with Mark's touchiness about taking the rap for delays in the Project. Everyone tends to blame these on hardware rather than software, although Mark has a point. A mistake in the program — allowing an errant electron to slip through a logic gate once in a million operations — can burn up your circuits as thoroughly as a bad solder joint. Yet, when presented with fried transistors, it's often hard to trace their sorry state back to a logical mishap. The joke is that no longer will problems with the system be identified as originating in either hardware or software. Instead, they will be referred to as "user irregularities."

I spend the afternoon working out with the betting practice box. Solenoids pop under the arch and heel of my foot as I receive a range of signals from one to nine. I translate them into patterns of numbers on the layout and scatter chips across the baize. I stand at the dining room table fielding buzz after buzz. I learn to distinguish the solenoids. I master the different frequencies. I memorize the numbers in each octant on the wheel. Bending over the baize with what becomes thoughtless precision, I concentrate on the task like a method actor intent on becoming the Marlon Brando of predictive roulette.

Doyne, still wearing the second pair of shoes in our set, walks into the room. "Let's do a range test," he says. Toggling microswitches with his toes, he simulates the clicking of mode and data switches during an actual game of roulette. For the first time the signals buzzing in my shoe are being transmitted from his computer to mine, exactly as they will be in Las Vegas. We run a dozen trials as Doyne walks progressively farther from me across the room.

"That was an eight," he says, referring to a midrange buzz on the back solenoid.

"Ditto," I confirm, getting the signal.

"Five."

"Five."

Another eight."

"No. I got a nine."

"That's the limit on our range. Beyond nine or ten feet, you're only going to get 'no-bet' signals." He turns on the betting practice box and watches me field numbers and toss chips onto the layout. "You're covering the patterns nicely," he observes, "but your technique is all wrong. As a big player, you'd better learn how to handle your chips."

He picks up a stack and covers numbers on the layout at twice my speed. "Instead of using two hands and dealing chips between your fingers like cards, you want to load them in one hand and drop them between your fingers like a coin dispenser. By not moving your wrist, you can get them out in half the time."

He steps back and watches me practice the new technique. "And there's one more thing. When you have everything else under control, relax. If all you're doing is gambling, you're supposed to be enjoying yourself."

Letty comes in through the front door. "Are you ready to go?" she asks, pulling a wad of bank notes out of her purse. "Here's the betting capital, twenty-five hundred dollars in cash. They were surprised at the bank when I walked in and asked for it. 'We usually don't hand out that much money without advance notice,' the manager told me. 'You'll have to believe me,' I said, 'but the circumstances are a bit unusual.'"

It had been agreed at the last Project meeting that we'd split up and drive to Las Vegas in two waves. Doyne and I are to leave momentarily in Letty's blue Fiat. As soon as they've finished a second set of shoes, Rob and Letty will follow in his Plymouth Duster. Mark intends to stay behind in Santa Cruz. He has several reasons, compounded out of pride and paranoia, for not wanting to go to Las Vegas. Either the computers work as he says they will, or they don't. Do we trust him? As for his paranoia, Letty has tried to convince him that the gambling statutes of Nevada do not explicitly forbid the carrying of predictive devices into casinos. Mark nonetheless pictures Mafiosi as big as Watusi tribesmen working him over in the back room of Caesars Palace, or suing him for every stick of furniture in his already modest house.

Toward evening I walk out the back gate to find Doyne inside the Blue Bus. The engine cover is off, and he is buried deep among the pistons with a variety of socket wrenches. Not running for the past month, the Bus had been parked in front of the house until the police came and threatened to tow it away. This being the first Las Vegas trip with no need of the Bus, Doyne had said he was going to push it into the barn. So I am surprised to find him now covered with grease and surrounded by engine parts. "What's going on?" I ask.

"It's a psychological thing," he says. "I don't know why, but I feel better when the Bus is running. I wouldn't want to leave town without it showing signs of life."

At dinner that night two items get passed around the table. The first is a clipping from the *San Francisco Chronicle*. Datelined Carson City, Nevada, it reads: "Gamblers lost $688.3 million in Nevada casinos during the three summer months, an increase of 8.1 percent from a year ago, which a state official says is pretty healthy in light of the recession."

"Who would think there were so many suckers in the world?" Letty asks.

The second item is a ten-page letter. "I'd like your comments on this," Doyne says. "I'm thinking of sending a copy to all the shareholders in Eudaemonic Enterprises. Now that we have a new generation of equipment, I think everyone with a slice of Pie should know what's going on."

Addressed "Dear Eudaemons," the letter begins: "Now that the christening of the ultimate pair of magic roulette shoes is imminent, the time is long past ripe to give an accounting of the status of the Project. Contained herein is a tentative division of the Eudaemonic Pie, together with a substantial revision and amplification of the original agreement. The enclosed photo of 'the sandwich' and 'the boat' should give you an idea of the current level of roulette technology. The sandwich shown is a complete data-taking computer, and the boat contains all the batteries, antennas, and two of three foot massagers (the other is on the computer)."

The letter describes our upcoming expedition "to the Nevada lettuce patch," where the new equipment is going to be tested during "a month of high-stakes playing." The letter also clarifies recent mutations in the Eudaemonic Pie, particularly the fact of its having gained a "front end," and proposes the following general division of the Pie, out of which will come Project members' individual slices.

Research and development (labor) 80%
Capital investment 19%
Seminal development 1%

"Before plunging into the details of the revised charter for Eudaemonia," the letter continues, "some of you may be interested in seeing the current tally of time and money that have been invested in roulette madness." There is a blank space on the page where Doyne meant to make this tally. Among all his other projects, he never got around to it. But if he had, the highlights of the Eudaemonic balance sheet would have read as follows: total capital investment of $15,000, the bulk of it coming from Doyne and Letty, with another few thousand dollars from Norman, Tom Ingerson, Dan Browne, and others. Of this money, $8500 was an advance on Mark's salary for a year and a half of overtime work. The remainder was spent on computer chips and other components. A separate fund for betting capital included as bankers for the Project Rob Shaw, Tom Ingerson, Letty, and Doyne's parents. But the truly astounding item on the balance sheet would have been the tally under "Goodwill." Heading the list of those donating labor and ideas during six years of roulette madness were Doyne, with thirty-five hundred hours of credit in the Eudaemonic Pie, Mark, with three thousand, and Norman, with two thousand.

Late that night I find Mark down in the Shop. His face is lit by the green glow of sine waves rolling over the oscilloscope. In front of him multicolored strands of bus wire stretch from the KIM computer to one of the sandwiches. "I'm testing the system," he says. "The KIM is running the sandwich through a cycle program. It works like a human being in pacing the computer through its predictions, except that the KIM makes no mistakes and never gets bored; it can produce the same result a thousand times in a row.

"I want to be sure the computer doesn't miss a step. I'm going to sit here awhile," he says. "Maybe all night."

In preparation for the yearly reunion of Pie holders, Eudaemons, physics wizards, and friends, work on transforming the house for its yearly fête begins early Halloween day. Out of the kitchen comes the smell of fruit pies and brownies. Doyne carries in a tank of nitrogen and demonstrates how, at −265°, the liquid can be made to skitter across the floor and bubble up into a cloud of steam. "You can also blow it out your mouth," he says, practicing what looks like fire-eating in reverse. "This should amuse all the physics wizards." Ingrid busies herself converting Norman's fish tanks into punch bowls filled with dry ice. Letty works on remaking the bedrooms into tactile and meditation chambers. Norman and Rob Shaw, who have borrowed a professional sound system with five-foot-high speakers, wire the living room into a quadraphonic disco with aluminum foil walls and a homemade light show of lasers and strobe lights. After scattering TV screens throughout the house, Jim Crutchfield turns the old Project Room into a production studio complete with makeup table, mirror, TV monitor, and video camera. One theme for this year's Halloween party is feedback. The camera and screens are being set up so that people thoughout the house will be able to watch themselves watching themselves.

In spite of the effort, the tone this year is subdued. There is a sense of impending breakup, with the house about to be sold and the residents scattered like seeds from a pod. Doyne to Los Alamos. Norman to Paris. Ingrid to San Rafael for a job with Lucas Films. And Letty to either New Mexico with Doyne or San Francisco to strike out on her own. Acknowledging the mood, the official title of this year's reunion is "The Last Halloween Party."

The house fills early with costumed dancers. The TV screens

wired to the camera in the front room play a nonstop show. A woman seated in front of the makeup mirror glues a mustache to her upper lip. A Phyllis Schlafly look-alike demonstrates how to cross and uncross her legs. A girl does jumping jacks. A monster eats a Boy Scout. Swirling across the screen and into the house are a procession of clowns and fairy princesses, a mirror with eye holes scratched in the glass, a Rubik's cube, woodland deities, Arab sheiks, and survivalists of various persuasions. I pick out Rob Lentz in a terry cloth burnoose. Jim Crutchfield floats by wearing purple tights and sunglasses. So convincing is the green putty bulging off her forehead like a hemorrhaging cerebrum that I barely recognize Lorna as Frankenstein. Lacquered in green tights and gold paint, Wendy dances with an ape wearing hairy safety goggles. It's Mark. Letty, in a black wig, sarong, and sandals, is dressed as a Balinese tourist. Norman appears in white shoes, black pants, a black shirt, a white tie, and a towering headdress of black and white crepe. "I'm basic integration," he announces. "As in mathematics?" I ask. "No," he answers. "As in black and white. It's the way of the future."

At midnight the house empties and everyone troops down the street to the parking lot of the New Riverside Szechuan restaurant. Here Rob Shaw, wearing a platinum wig and long johns, puts on a show of homemade pyrotechnics. The crowd exclaims over the rockets and Roman candles bursting overhead: "Score one for Nicaragua. A little farther to the east, Rob. Maybe we can take out Washington."

Back inside the house, the music throbs as dancers whirl under laser beams and mirror balls. Two fairy princesses dance cheek to cheek. A Jesuit priest with fangs makes out with a bearded nun. The third sex is much in evidence, but there seems to be a cultural split this year between decadence and punk — the sex-change–transvestite look and your basic-black nihilism. Ingrid, wearing motorcycle boots, a cut-off T-shirt, and a Vaseline duck's ass hairdo has come as a Hell's Angel. The front of her shirt says "Mustache Rides," and the back reads:

> Born on a mountain
> Raised in a cave
> Biking and sex
> Is all I crave

Doyne, inclining toward the decadent end of the spectrum, is dressed in stockings and heels. He wears red lipstick, bangles, gold earrings, a corset, and a brassiere with a wad of play money stuffed into his bosom. Swimming over his head is a blond wig, which gives him the ratted, floosy look of a Las Vegas call girl long since gone to seed.

The house shakes to the sound of Xene singing "Johnny Hit and Run Pauline." Liquid nitrogen steams up off the floors. The TV monitors begin to whirl and pulse with strange patterns. I discover Ralph Abraham with the video camera in his hand, and he's pointing it straight at one of the TV screens. "It's a feedback loop," he tells me, motioning toward the balls of light pulsating on the screen. "The camera shoots an image of itself shooting an image of itself in endless regression. Because of a split-second delay in its focusing, the image is unstable. So you get *continuous* feedback. It's a kind of sensory overload." The pulsating balls and crescents of light perform a kaleidoscopic dance of electrons. The screen glows luminous with patterns that never repeat themselves. "The system is so overloaded that it can't resolve itself into a stable pattern," says Ralph. "It's a perfect example of strange attraction."

Late the next morning, nursing our hangovers over crab salad served on the sun deck at Aldo's, we blink out at the blue waves and yachts bobbing in the harbor. Mark picks at his food and avoids looking at the sun. Ingrid excuses herself and walks off the porch to lie in a bed of ice plant. "I think the party was a throwback to the old days," says Norman. "It was decadent rather than punk, and decadence is out of fashion."

"I had the sense of an era coming to an end," Doyne says, "the feeling that this may really be the last party. Anyway," he adds, turning toward me, "it was a good sendoff. Let's pack our bags and get out of town."

16

Cleopatra's Barge

Desperate, but not serious
Adam and the Ants

That afternoon Doyne and I load two pairs of magic shoes and socks into the Fiat and head for Las Vegas. We also take with us twenty-five hundred dollars in cash and a nice selection of stay-pressed pants and Hawaiian floral-print shirts. Driving south on Route 101 through the vineyards and grazing country above Monterey Bay Doyne turns to me. "Do you know why we called the computer sandwich a sandwich?"

"No."

"Mark figured that if we ever got in a really tight spot we could eat it. One of the features we thought of including in the shoe was built-in ketchup packets. But we expected we'd run into problems tweaking up the ketchup-packet holders."

Doyne rummages through a paper bag and pulls out a brownie left over from the party. "I wish I were in a casino right now. Which isn't to say that I'm not looking forward to Bakersfield."

At Paso Robles we turn east into the foothills of the Diablo Range, on the other side of which lies Bakersfield and a steeper climb into the Sierra Nevada. Not until crossing Tehachapi Pass will we drop down into Boron, Barstow, Baker, and the flat run through the Mojave Desert to Las Vegas.

The car suddenly swerves. Doyne cranes his neck to stare out the back window. "Did you see that tarantula?" he yells.

"I didn't see anything," I reply, turning to look at the pavement receding behind us.

"It was as big as a crab. It was huge. Or maybe it was a bat, or a vampire," he says, as we both start laughing.

On top of the pass through the Diablos we look east over the Central Valley to the high Sierra. A dusting of snow on their peaks glows red with light from the setting sun. We cross Highway 5, the major north-south truck route, and find ourselves driving through a featureless lowland scratched into row upon row of cotton fields. Outside of Wasco, where it turns into fruit and nut country, we find a surprising number of '57 Chevies cruising the highway. A banner over Main Street advertises the Wasco Turkey Shoot, and we hear the sound of rifles popping in the fields.

"Our Explorer Post put on a turkey shoot once," Doyne tells me. "We were trying to raise money for a trip to South America."

Dan Hicks and His Hot Licks come on the radio playing their special brand of Hawaiian tutti-frutti music. "If I ever make any money," Doyne says, "I'd like to get a stereo and buy a few records. I'd start with early jazz, the vintage stuff from the late twenties and thirties — Louis Armstrong, Nat King Cole, Django Reinhardt, Stéphane Grappelli, Mike Lovell France. They could take anything from 'Sweet Sue' to 'Sewanee River' and turn it into music you could dance to. Then I'd pick up some Fats Waller, Willie Maybaum, and other people who played blues piano in the early fifties. And the Boswell Sisters. They sang swing and jazz that always hopped. I'd buy all of Hank Williams and Dan Hicks, at least his early records. I'd want a decent amount of Cream, for whenever I got that Creamy feeling, and the early Beatles and Stones, along with the Kinks and Buffalo Springfield. Add to that list the Coasters — definitely the Coasters. And I almost forgot the complete Chuck Berry."

Tehachapi Pass is OPEN, say the flashing road signs. We line up behind the big rigs and file over the mountain in a haze of diesel. Coasting down into the town of Mojave, we pass the Bel Aire Motel and a neon sign lit over a palm-reading salon. "I'd stop and go in for a reading," Doyne says, "but now that we're in the desert we should cruise."

Transparent night settles around us, with stars shining in it like gems pushed forward on a jeweler's mat. The moon rises gibbous. Emptying the last paper bag of apples and brownies, we pass through a buzz of neon at Stateline before dropping back into the blue envelope of desert night.

Forty miles out of Las Vegas the sky lightens. Swept up in a stream of traffic, we roll toward a preternatural dawn that turns from burnt umber to pink until suddenly, down below us in a valley stretching east from the Spring Mountains, we come on the full glow of this pleasure dome shining in the desert. Giant clusters of light erupt and mutate in what looks like a time-lapse movie of flowers blooming in neon. Silver veins of light stretch far into the desert as the city-organism below us winks and spins in photokinesis.

We exit off the interstate onto Las Vegas Boulevard South, otherwise known as the Strip. The traffic rolls slowly from one burst of color to the next. Spotlit in this neon garden are the fountains, plaster statuary, Persian tiles, Roman arches, porticos, and loggias that decorate casinos ranging in style, as Robert Venturi puts it, from Miami Moroccan and Hollywood Orgasmic to Niemeyer Moorish, Roman Orgiastic, Arabian Tudor, and Bauhaus Hawaiian. A frenzy of light exploding in starbursts and whirling over aluminum palm trees illuminates the huge, seven-story signs that announce the evening's entertainment. "Wayne Newton is playing tonight at the Aladdin," Doyne intones with the mock enthusiasm of a tour conductor. "The Aladdin used to be a dump. But they've remodeled it with a neon sign as big as everybody else's. Things change.

"Here's the MGM Grand, rebuilt after the fire, and the Jockey Club, a new casino. On your right we have the Barbary Coast and Maxim and the Flamingo Hotel, where Bugsy Siegel started the whole shebang. On your left we have the triumphal entrance to Caesars Palace. The elevated conveyer belt allows you to drop into the casino from the heavens. The statuary, as you'll notice, is anatomically enhanced." The neon sign at Caesars Palace, which is decorated with statues of centurions and steam bath towel boys, announces that Cher is playing the Circus Maximus, while Pupi Campo and Bruce Westcott are rocking out on Cleopatra's Barge.

"We're approaching the Wild World of Burlesque at the Holiday Casino, and off to our right is the Imperial Palace, for Oriental pleasures. We're bedazzled in front of the Nob Hill Casino, the Sands, and the Castaways, which is another of the town's new attractions." Teenagers in low-riders grind their gears and honk as we cruise together past the Frontier, the Desert Inn, and the Stardust.

"There's the Silver Slipper, famous for its ninty-nine-cent break-fast. Although I see on the sign that it's gone up to a dollar twenty-nine. And here's the Silver City Casino, the scene of our first big win, where Ingrid cleared five hundred dollars in thirty minutes of play. On your right is the Landmark Tower, and coming up on the left, for family gambling, we have Circus Circus. As you know, that casino has been the scene of many a successful roulette session. Next door at the Stardust we have for entertainment 'The All New Direct from Paris Lido Show Les Bluebells Girls with a Cast of a Hundred.'"

Farther down the Strip, a picture of Loretta Lynn lights up the front of the Riviera. "She looks nice," says Doyne. "We should check her out." Past the Silverbird and the Candle Light Wedding Chapel, advertising "Immediate Wedding Services All Checks OK," we turn right at Foxy's Firehouse Casino onto Sahara Avenue, and then take a quick left onto Paradise Road.

"We've stayed here before," Doyne says, stopping at the Brooks Motel. "You can't beat the location, and it's cheap." The manager takes a week's rent in advance, and we drive around back to unload the car. We hear the sound of gunshots and the neighing of horses as TV screens flicker in the windows. Entering a small courtyard filled by a swimming pool and a palm tree, we walk up a flight of stairs to our apartment, which has one bedroom and a living room divided from the kitchen by a low partition. A sliding door opens onto a balcony overlooking the pool.

"That's where the second-story men get in," Doyne warns me, nodding toward the balcony. "In Las Vegas there's a lot of funny money floating around in people's pockets, and these guys figure they're helping to keep it in circulation. So if you open the window at night, you might want to sleep with your shoes on." I wonder, is he referring to shoes with or without computers in them?

It's one in the morning, and we're tired. But we're also jazzed by being in Las Vegas. So we lock up the apartment and drive to Fremont Street. Here we find the three blocks of casinos that constitute what Las Vegas calls downtown. We split up and stroll the neon corridor. Sauntering into the Mint and the Golden Nugget, I stop to watch the action at the roulette wheels. In a small notebook — official issue of the Project — I record wheel data on high sides and rotor speeds. I jot notes on the croupiers and map the layout of tables on the floor. "Las Vegas wheels are as tilted as ever," Doyne declares when we meet back at the car. "I'd say it's looking good. Very good."

We wake up late the next morning. The day is already hot with dry air wrapped around us like a garment bag. I step out onto the balcony. Below me the manager is filling the Coke machine next to the pool with red and white cans. Last night when we checked in she had showed us a picture of Melvin Dumar. Signed "To the Brooks Motel with fond regards," the photograph showed a man with a pompadour hairdo and a pout on his face. "Mr. Dumar is heir to the Howard Hughes fortune," said the manager, "but at the present time he is doing an Elvis act. When he's in town, Mr. Dumar always stays at the Brooks."

I look out from the balcony toward the mountains that rim Las Vegas. The Spring Range rises to the west and Sunrise Peak to the east. What I can see of Las Vegas itself includes the roof of Foxy's Firehouse Casino, where a large sign pops on and off to advertise FREE HAMBURGERS   FREE DRINKS   NO LIMIT. Across from Foxy's is the tower block leaning over the pool at the Sahara Hotel, while to my left the view takes in Paradise Road and the tops of women's heads whose hair is dark at the roots but golden blond by the time it flips over their shoulders. The women peer from behind sunglasses at a choice of storefronts that includes an acupuncturist, an abortion clinic, and a specialist in cosmetic breast surgery and silicone implants.

Doyne and I drive down to the Golden Gate for a breakfast of pale eggs jiggling in bacon grease. We take another stroll down Fremont Street and stop at the Golden Nugget to watch them leveling a roulette wheel. A croupier studies the bubble floating in a spirit level laid over the wheel. A security guard lifts the wheel housing, and the croupier stoops to adjust tumblers on the feet of the table.

"You can't reliably level a wheel that way," Doyne tells me later. "I'm surprised at how crudely they do it."

Spending the afternoon at the motel, I spread the layout and chips over our coffee table and put on a pair of magic shoes to work out with the betting practice box. Doyne tweaks up his solenoid plungers and cuts holes in his socks. He drives out to Radio Shack

for batteries and takes a nap. After a dinner of microwaved tortillas at Carlos Murphy's Irish Mexican Café, we head for Fremont
Street and our first night's work.

We drive into the parking garage at Benny Binion's Horseshoe
Club and circle up the ramp to the third floor. We park the Fiat
and change into our gambling shoes. Stepping away from the car,
Doyne clicks a range of data into his computer. A second later I get
a prediction buzzed into my shoe.

"What was that?" he asks.

"A three."

"Right. And this one?"

"A nine."

"O.K. And this one?"

"A five. Maybe a six."

"That's strange. I got a nine. We should be transmitting ten feet
from shoe to shoe. But the bad signals mean we're not doing any
better than six feet. You'll have to stick close to me along the
layout."

Doyne hands me a roll of hundred-dollar bills and walks to the
elevator. I wait five minutes before following him down into Glitter Gulch. After strolling the casinos and noting the action, I reach
the Sundance a half hour later. I skirt the gambling floor and head
for the rear of the casino. From there I watch Doyne standing at
one of the two roulette wheels in play. Giving him time to finish
setting parameters on the computer, I walk past the table, and on
my second tour he places a side bet on the even numbers: my signal to buy into the game.

Seated to my left is a man wearing a string tie and a Stetson. To
my right is a Filipino smoking a cigar. "Let's see how Lady Luck is
doing tonight," I say, placing my first bet. "It sure is a warm night
for November," complains the man in the hat.

I pick up a solenoid buzz and easily cover the numbers on the
layout. I chat about the weather and then suffer a sweet rush as
my chosen octant on the wheel is hit dead-center. With a payout
at thirty-five to one, this is a tender moment. I feel myself getting
big with consumptive excess. I imagine money dribbling from my
pockets as I give myself over to the delights of fast horses and
women, Caribbean hideaways, duck hunting in the Urals, balloon
excursions with Malcolm Forbes. There will be more than enough

left over to support all the good causes in the world. I intend to be the rare example of a nice guy who gets rich and stays nice. Beating roulette isn't going to go to my head, just my pocket. As the croupier rakes a stack of chips toward me across the table, I turn and order a drink from the hostess.

But then I notice something strange about the computer's signals. There seems to be a problem with the solenoids. Every few seconds, apparently at random, they pop off with different vibrations. I start to place a bet with one signal and then switch midway on getting another. Or I wait for a buzz and get nothing. Thinking my receiver might be out of range, I move closer to Doyne. While trying to distinguish the good signals from the bad, I find myself tossing out chips at random to cover my confusion.

The computer underfoot tinks and whirs from one prediction to the next. Extraneous buzzes pop off when the ball isn't even in motion. They follow rapid-fire, one on top of another. Some are clean and readable. Others feel like accidents — little mechanical stutters from a computer embarrassed at how badly it's performing. With more and more buzzes coming out of nowhere, my shoe feels like a foot massager run amuck. I'm getting a ten-week course in acupuncture all in one night. I place bets indiscriminately and wait for a sign from Doyne. Mr. String Tie has been cleaned out, and the Filipino is losing his touch. Sucking hard on his cigar, he scatters chips across the layout. A pile of them straight up on number 17 tumbles over and has to be straightened up by the croupier. Standing next to me at the wheel, his face tightened into an edgy frown, Doyne bets the house minimum on odd or even numbers.

I am doing my best to make sense out of the Chinese foot massage. So as not to quail before the law of probability, which allows for the remote chance that a bettor with a 40 percent advantage over the house can still get wiped out, I have been told to play roulette until given the sign to quit. I am headed for the cleaners by the time Doyne places a chip on 00. "It just doesn't feel like my lucky night," I tell the croupier. He claps his hands and the pit boss come over to watch my roulette chips get converted back into casino currency.

I walk to the cashier's cage and then out onto Fremont Street. Meeting me later at the Golden Nugget, Doyne joins me in a booth at the back of the coffee shop. His face is gray with fatigue. "My computer was crashing left and right," he says, "but what's really killing us is random buzzes. We're getting swamped by spurious noise." *Noise* is the term in electronics for a signal with no function. Over the next few days, I am going to hear a lot about noise.

I wake late the next morning and walk out onto our balcony. Across the central oasis, with its swimming pool and palm tree, I look into the facing apartment to see a man wearing a stocking cap. He sits in a chair tilted against the wall and smokes a cigarette. A woman in a bathrobe serves him a plate of what look like scrambled eggs with ketchup. Below us, the manager is fishing soda cans out of the swimming pool with a net. The day is bright and dry. Too dry. Even in November the desert could desiccate you.

On Paradise Road the signs are lit in front of the abortion clinic, the acupuncturist, and the specialist in cosmetic breast surgery. Women park their cars behind the buildings and pause for a moment before stepping out. In the opposite direction, past the neon humming over Foxy's Firehouse Casino, lie avenues and housing projects laid over the desert like a printed circuit. Out beyond the copper bowl of the city stretch the red hills of the Las Vegas Range.

I leave the curtains closed and step back inside our two-room apartment. It comes complete with kitchen and dining nook, although the dining nook table is now buried under a pile of computer sandwiches, battery boats, alligator clips, ohmmeters, solder, and electrical tape. Scattered over the rest of the apartment are wiring diagrams, data manuals, and shoes with their insoles flopping out. Doyne lies supine in a sleeping bag on the living room floor. This is a prophylactic measure for a bad back. Down feathers cling to his hair as he sticks his head out of the bag and yawns. His face, rumpled and out of true, is not yet gathered into cognition.

"I had the weirdest dream," he says. "I was inside a casino, one of the large ones, maybe the MGM Grand or Caesars Palace. But it was also a church, with candles and incense burning and Gregorian chants coming over the loudspeakers. There were nuns and priests officiating at what looked like a roomful of altars. Everyone was praying, and it seemed to be a very religious and holy place. But then when you got up close you saw that the nuns had bare legs. The priests were really croupiers, and the worshipers at these altars were blackjack and roulette players taking the Holy Sacrament in the form of casino chips and Bloody Marys."

After breakfast, Doyne calls Santa Cruz for a consultation with Mark. We have no phone in the apartment, so he uses the pay phone in the courtyard next to the swimming pool. A long conversation is interrupted for continuity tests and other electronic probes into the hardware. The tests produce no solution. Worse yet, they discover no problem. Random errors in computers that sometimes work and sometimes don't are the hardest of all to troubleshoot. There is no way to shake a bug out of a computer until it manifests itself.

Doyne hikes upstairs from his last call to California. "Mark wants us to do a reality check. He thinks we've gone weird on him out in the desert. We're supposed to go back into the casinos and run a range test."

We load our shoes with computers and batteries and drive around the block to the Strip. Here great white wings flicker over the Silverbird. The nose on a huge clown blinks on and off at Circus Circus. A firmament of light shoots above the Stardust. A massive red *R* beckons at the Riviera. As we drive down the Boulevard, an afternoon sandstorm whips over the waste lots. The air silts up with dust. Bundles of tumbleweed blow across the road, and the hookers working the street take shelter behind the signposts. After turning into the parking lot at the Stardust, Doyne hands me a plastic bag into which he has stuffed the betting practice box. "Give me a five-minute lead," he orders. "Then switch on the box and follow me into the casino."

Designed originally to hold canceled checks, the box looks inconspicuous from the outside. But on peering under the lid one discovers a small computer, bundles of batteries, a radio transmitter, and a light-emitting diode flashing numbers from one to nine. Engineered to blast out solenoid buzzes at random, the box is now getting pressed into service as a portable transmitter. I am supposed to follow Doyne around the casino while he compares the strong signals coming from the box to those generated by the computer sandwich in his shoe.

"In case anyone asks," I inquire, "what exactly am I carrying here?" An extortion ring had recently blown up Harvey's Casino in Lake Tahoe, and the bomb, disguised as a computer, had been detonated by radio signals.

"You could say you're making a movie. This is a remote control device for your camera." Doyne is already out the door before I can remind him that movie making is also forbidden in casinos.

I wait five minutes before switching on the box. On entering the Stardust I pause to let my eyes adjust to the light. Except for a general layering of plush and twinkling of light bulbs up in the rafters, this cavernous hall could double as the Cleveland convention center or part of the Newark airport. The only action at the tables comes from a few diehards. I spot Doyne at the far end of the room and make a slow pass in front of him. As he walks under the Eye in the Sky, I shadow him around the floor.

"The signals were coming in all right," he tells me back in the car. "But the range was good only up to six or seven feet."

Driving south on the Strip toward Caesars Palace and the really big casinos, we turn into the parking lot at the Silver Slipper. "I'm going back in," Doyne says. "This time I'll transmit out of the shoe and hold the receiving computer in my hand." He wraps my computer sandwich in a plastic case borrowed from the Project's tape recorder and puts it to his ear. "What do you think?" he asks. "Does it look like I'm listening to the radio, or something like that?"

He loads the transmitting computer and fresh batteries into his shoe, bucks the wind on the tarmac, and disappears through the front door of the Silver Slipper. I follow a few minutes later, again carrying the betting practice box in a plastic sack. Doyne and I are engaged in the process of experimental physics. When theory fails to produce an answer, a scientist has no choice but to head for the field and gather data. I stop near the entrance to watch the race results go up on the big board, and then walk to the casino floor.

I find Doyne standing in front of the craps tables with the computer held to his ear. Wearing blue jeans and a striped cotton shirt, he looks like a farmboy dressed for a visit to the local Sears catalogue store. But if this is a radio he's holding, it's odd that no sound is coming out of it. There's no finger clicking, gum chewing, bebop, or sign of anything on Doyne's face other than total stupefaction. This means he's driving around the mode map with his toes. But I'm not the only one who thinks he looks suspicious.

The shift boss stands elevated on a podium behind the craps tables. A red light flashes on his telephone. Pit bosses from all over the floor look to his desk. A half dozen men in brown suits converge on Doyne, who by now has joined me in a record-breaking strut

toward the door. We run to the car and lay a patch out the driveway as the brown suits hit the parking lot.

"Luckily we have enough to spare," says Doyne, "but that's one
casino we definitely won't be playing."

"There are two ways to proceed," Doyne explains as he plugs the
soldering gun into the kitchen socket. "We can spend another couple of years testing the computer for sources of noise, or we can try
for a quick fix. Mark suggests I look at the top of the sandwich and
resolder the on-off line. It's a bolt-tightening operation, like checking to see if your steering wheel is connected to the car. Then I'm
going to solder another capacitor into the circuitry of the receiving
computer. That should boost its power and make the solenoid
buzzes stronger. We worried a lot in the past about keeping them
quiet, but at the moment all I care about is blasting out a signal.
Why worry about the Mafia listening in on your shoes when you
can't even play roulette?"

Doyne examines the computer sandwich in front of him and
looks for the little cul-de-sac of copper that constitutes the on-off
circuit. Under the solder points on this section of the PC board lies
the microprocessor, for which the on-off line acts as a gate governing access to the CPU. "I hate this part more than anything," he
mutters, poking the sandwich with an ohmmeter. "I seem to spend
most of my time on these trips troubleshooting equipment. There
are usually four or five people sitting around waiting for me to pull
off the miraculous fix. This time at least it's just the two of us."

I sit on the couch, reading back issues of *Gambling Times*. A TV
hangs in front of me on a rack bolted to the wall. "News flash," a
man comes on the screen to announce. "William Holden, the actor
who played the all-American good guy, has just died. The romantic
hero of *Bridge on the River Kwai* and *Sunset Boulevard*, Holden was
best man at the 1952 wedding of Nancy and Ronald Reagan. President Reagan, when informed of his friend's death, said he was
shocked and has a great sense of grief."

I stand up and change channels through a soap opera, an *I Love
Lucy* rerun, and a PBS special, "The Expanding Universe." While
spinning the dial I notice something strange. On the coffee table
between the couch and the TV sit one of our two computers and a
battery boat. As I flip the channel selector, the solenoids start pop-

ping like Mexican jumping beans. I squelch the volume and whirl the dial through TV snow. I call Doyne over, and together we stare at the twitching solenoids.

"That's it," he says. "You found the problem. The Eye in the Sky is nothing but a giant TV installation. No wonder we're getting killed by noise. It's like the Russians jamming the Voice of America. They throw out so many signals that only garbage gets through."

The casinos are a swamp of electronic noise. It comes from the surveillance systems, but also from low-frequency radiation given off by the neon signs and slot machines. As defined by Claude Shannon, information is the amount of surprise in a system. Noise, in this case, is an unpleasant amount of surprise, which is why Shannon chose to measure it in terms of entropy. As in Grand Central Station at rush hour, when the face you search to retrieve from those standing under the clock refuses to materialize, noise is the too-muchness of everything happening all at once. Noise is the audible trace of a system slipping from order into chaos. Bad trips, static, psychoses, and the random buzzing of solenoids are all examples of excess information.

"We can change our radio frequency and retune the equipment," Doyne says. "The computer's already designed to float above the ambient noise. We'll just have to float it higher."

Resoldered and retuned, the sandwiches get loaded into our shoes by late afternoon. We drive out for dinner and then head downtown. Gripping the steering wheel with one hand, Doyne uses the other to hold a shoe up to his ear. "I want to listen to the solenoids. I wonder if they'll start popping in front of the neon." As we slow at the intersection of Las Vegas Boulevard South and Flamingo Road, a black man in a Cadillac pulls alongside. He lowers his window and points to the shoe. "I can hear it, man," he calls to Doyne. "It sure do got a funky beat." The man laughs uproariously and drives off snapping his fingers.

On a chilly night in November, before the gamblers arrive for the holidays, business is slow in Glitter Gulch. The neon whines overhead and the barkers on the pavement seem to be hustling each other. I walk from the Mint to the Union Plaza, which straddles Fremont Street and caps the far end of the Gulch. I watch the wheels in play and jot notes on their tilt. At the Golden Gate, I push through the crowd around the craps tables to find Doyne facing the single roulette wheel in play. He places a bet on red, my signal to take a five-minute walk.

Sitting in front of the keno board, I chew on my pencil and pretend to fill out betting forms. I know things are bad when the solenoids in my shoe start popping off at random. Spurious noise is back for a visit. Whenever I drift over to the roulette wheel, Doyne places a bet on red. It turns into a long night of five-minute walks.

"We have a new problem," he says, meeting me later in the coffee shop at the Golden Nugget. "My computer keeps crashing. I get warning buzzes on the solenoids, a lot of nines, and then the system goes dead. It seems to be powering up and down at will, as if it's getting lost in its program and doesn't know where to go. Finally, it died completely. I've replaced all the batteries and checked the connections, but no go. It's just sitting in my shoe, not doing anything."

Doyne loks pale. His fingers curl and uncurl a corner of the placemat. "It's only midnight," I say, trying to cheer him up. "You want to do something fun? Like go gambling? We could play the slot machines at the Jolly Trolly and win a free hamburger. Or we could catch the last show at the Lido de Paris and finish up on the Strip with a dollar-twenty-nine steak-and-egg breakfast."

Instead of playing the slots at the Jolly Trolly, we phone Len Zane. "We need an oscilloscope," Doyne tells him. "Can you help us out?" As head of the physics department at the University of Nevada, Zane is fond of Doyne and tinkerers in general. "Come by the house," he says, "and I'll see what I can do for you."

Zane sets us up for the night at a workbench complete with solder guns, AC-DC multivoltage power supplies, and a large Tektronix oscilloscope covered with dials. Doyne plugs two needle-nosed probes into the scope and hesitates a moment over the sandwich lying in front of him. With its chips packed inside, the only visible parts of the computer are the backsides of its PC boards, which are covered with nubbins of solder wherever a chip has been stuck. The solder points shine like tin roofs over the silicon sheds hidden beneath them, and these silvery dots are Doyne's sole guide to what lies inside the computer. Only by probing for sine waves can one diagnose whether there is life below.

This delicate operation is comparable to a brain scan. Short of melting the wax out of it, there is no way to open up the computer sandwich, just as there is no simple way to lift off the top of the skull. All one can do is search for electronic discontinuities. Spikes appearing in an otherwise regular wave that indicate a loose connection or a burnt-out chip. The examination of a computer from the outside, pin by pin, is tedious work. It also calls for the hands of a surgeon, because a slip of the electronic probe can itself accidentally blow out a component.

In the small hours of the morning Doyne discovers a discontinuity in one of the lines. He resolders the pin and then all hell breaks loose. Where formerly there had been a clean wave on the scope, there now flashes across its screen a crosscurrent of peaks and troughs. Doyne reburns the solder joint and pokes his needles from one part of the sandwich to another. Finding nothing but a roaring sea of noise, he phones Mark in California, describes the problem, and tells him to call back when he's thought about it. Throughout the night, with Mark simulating the problem back in Santa Cruz, the two of them speculate about "the fifty-six-ohm resistor being open circuited, the on-line going high, or noise getting kicked into the amplifier," until Doyne is finally forced to admit, "There's chaos everywhere."

He hangs up the phone and turns to me. "This is the Polish fix. I'm talking to the only person who knows how to get this computer running, and he's five hundred miles away." The first shift is already on the highways heading for work when Doyne places a final call to California. "I give up," he tells Mark. "I want you to wax the second set of computers and get them out here as fast as possible with Letty and Rob. If we're going to work, we need some tools."

Doyne and I leave the shop at dawn. After a cold night in the desert, with temperatures down in the thirties, the mountains are covered with wispy clouds that won't burn off until afternoon. By then the temperature will have crept up into the sixties, and the wind will have risen to blow the dust off the bare lots between the casinos and condominiums. The weather is all we have to think about as we wait for a take-out order of computer sandwiches to reach us from five hundred miles across the desert.

Cooling my heels waiting for the second team to arrive from California, I tour downtown casinos, eyeball the wheels, map layouts, and return to the motel to find Doyne where I left him, sitting at the kitchen table poking an ohmmeter into a computer sandwich. I walk out again for an afternoon promenade on the Strip. Perpendicular to the high-rise signs on Las Vegas Boulevard South branch smaller arteries of plastic that run to gravel out in the desert, where the sun is setting neon red behind the Spring Range. The mountains on the horizon offer a lunar calm, but down in the conduits of Las Vegas the air is charged with megawatts. While tubes of light flash around me in the shape of boomerangs, star bursts, and intertidal organisms, sunset on the Strip is one throbbing light show superimposed on top of another.

As mercenary as it may be, Las Vegas is also a mystery, or at least a set of paradoxes. Casino gambling was once the sport of kings and aristocrats, and the genius of Las Vegas lies in having elevated everyone to the peerage. Caesars Palace is open to the public, and anyone can be sheik for a day at the Sahara. But leisure here is a mirage. a calculated feat of social engineering. The casinos offer perfectly controlled environments that pretend to be free and at risk, but actually everything in them, from gambling to sex, is geared into a machine for maximizing profit.

Absent of commodities in the traditional sense — things like pork bellies and wing nuts — Las Vegas itself has become a commodity. To do so, the city had to transform itself into a fetish and phantasm of pleasure. Among fetishes in the modern world, Las Vegas is one of the most potent. Who, when polled, fails to associate it with pleasure in excess? Las Vegas is a dreamscape, a simulacrum not to be missed by philosophers interested in studying the paradox of false pleasure. This city of signs and symbols — this green world of semiotics — teems with what might be called the "fun cue." This is the sign that denotes the *idea* of fun, and it appears in the form of towel boys standing solicitously at the door of the steam room, or as hostesses with their bottoms pulled tight into leotards. The linguistic correlate for the fun cue — the word most often employed to denote the idea of fun — is *free*, as in free drinks, free breakfast, free champagne, free hamburgers, free tourist gambling packages. When you know what to look for, the fun cue can be seen everywhere in Las Vegas: in bars decorated with palm trees, in motel lobbies with birds in rattan cages, or in the welcoming ease with which croupiers bend forward to convert your money into chips.

*

Stiff from having crossed the desert at night in a car with no heat, Letty and Rob arrive early the next morning. Letty carries two computer sandwiches into the apartment. "The B team to the rescue," she says with a smile.

Rumpled but hearty, Rob follows with a portable oscilloscope and toolbox. He greets Doyne with a bear hug. "Where's the action?" he asks. "I'm looking forward to some wild and crazy times." Then he announces in a deadpan voice, "The new computers test out perfectly. We checked them thoroughly for five minutes before getting in the car."

"It took Mark a few extra hours to wax the components," Letty reports. "Every time you called him on the phone, he'd take them out of the oven for fear of overcooking them. Then there was a big scare when he thought he'd assembled the sandwich upside down. It turned out that he hadn't. But the paranoia was thick while he checked it out."

"It's great to see you guys," Doyne says. "Why don't you get some sleep, and then later tonight we can play roulette."

But instead of going to sleep, Rob heads for the kitchen table. "What seems to be the problem here?" he asks, sitting down in front of our two dysfunctional computers. "I understand your sandwiches are suffering from spurious noise. Is that right? We don't usually make house calls. But you told us the situation was desperate."

He snaps the cover off the oscilloscope and holds the needles in his large hands. As he probes the computers, patch cords dangle from his mouth. He mumbles to Doyne about discontinuities in the address line. "It's hard to understand how this could happen, unless your PIA has burned out." He plugs in the solder gun and waits for it to heat up. Doyne sits next to him, studying wiring diagrams. "Run it by me again," Rob says. "What's supposed to happen in the power-up, power-down sequence?"

By the end of the day, sandwiches and boats cover the table like war casualties lined up for triage. Doyne scrutinizes pin-out maps describing where in the microcrystalline wax the unseen chips are buried. The oscilloscope burns green in front of him with glitches. Smoke hangs in the air, along with the bitter smell of solder.

Doyne and Rob run range tests and probes. They cut lines and retune components. They tweak and solder, and entire days go by in which the boredom is punctuated only by false alerts in which everyone scrambles to tie on magic shoes, only to find these sessions time and again aborted by attacks of spurious noise. Solenoids pop off. Computers motor around their programs at random, get lost, and burn up batteries. The machines tink and buzz with increasing weakness in their vital signs. Most depressing of all is the news that the new computers are no better off than the old ones. As if having contracted a contagious disease, they too are now afflicted with spuriosity. Theories and rumors multiply as fast as glitches on the scope. Has a bad boat burned out all the computers? Was there a flaw in the design, like something wrong with the on-off line? Or is the environment in Las Vegas itself just too hostile?

"Maybe we were overly strict about the design requirements," says Doyne, who is slowly resigning himself to becoming philosophic. "It might be asking too much to put a computer in a shoe. There are so many different things in there that can screw each other up. Let's face it, building a computer to walk on is a difficult problem to solve."

Viewing the current dilemma as a temporary setback, he speculates on what the Project might do next. "By thinking about it, we should be able to get rid of whatever it is that's making the system flaky. I see exactly what the next generation of equipment would look like if we worked backwards and de-evolutionized the design. I'd take the computer out of the shoe and strap it onto my leg, along with a little garter belt for holding the solenoids. I'd put the battery pack on the other side of the computer, or down in the solenoid garter belt. Then I'd fill up a shoe with a hefty antenna and toe switch. There's no reason not to keep the mode transmitter just like it is in the other shoe. We haven't had any problems with that. Unlike the old days, when we had cables strung from our toes to our armpits, the new system would have a single power harness running from the shoe to thigh level. Building a new computer isn't that big a deal. All I imagine doing is spreading the system up the leg."

"God," Letty exclaims, "it sounds like the Project is starting up all over again. Is this the first step down the old road?"

"You're right," Doyne admits. "It may be a waste of time thinking about the next generation of roulette computers. For all I know, the Project is dead."

"What do you want to do about it?"

"We have three alternatives," Doyne says. "We could build a new generation of equipment by plugging away at it in our spare time."

"That's a bad idea."

"O.K. It's a bad idea. There are two other things we could do. We could find investors and hire a professional technician out of the Silicon Valley. For twenty-five thousand dollars we could get it done right."

"That means more people expecting a bite out of the Eudaemonic Pie."

"The third thing we could do is quit. We already have our statistical victory. We've proved we can beat the casinos with a large advantage. So we forget about pushing the stakes and making a lot of money and all the rest of it. We just snap a few photographs and show them to our grandchildren."

We are sitting in front of the TV on a Saturday night eating "homemade" Betty Crocker date bars and watching Peter Ustinov narrate a PBS program called "Einstein's Universe."

"Hey, Al," Doyne calls to the image on the TV screen. "What would you do if you were us?"

Rob wanders over to the kitchen table and takes a final poke at the computer. "My latest theory is that we have a problem with the RAM."

"That's the only thing we haven't slung any mud at yet," Doyne says, "and we might as well."

"It's not the CPU. It's not the PIA. It's not the EPROM."

"It's not the CIA," Letty quips. "It's not the NRC."

"Maybe it's the FBI," Doyne says. Picking up Rob's guitar, he strums a few chords and starts singing "Me and My Uncle" in his best New Mexico twang.

"Hey, you guys," Letty says. "It's Saturday night. We're supposed to be out on the town having fun."

"That's right," Rob agrees. "I'm too young to turn into a nerd."

"We know all the high spots, don't we?" Doyne says, turning toward me. "What do you say we take everybody out and show 'em a good time?"

"Let's get dressed up," Letty proposes. "I want to do the whole thing right. Put on my shoes and load them with a computer and batteries. I don't care if they're not working. I want to experience walking around Las Vegas with a computer in my shoe. Just this once I want to head out the door and feel what it's like to be powered up and ready to play."

"That's what we should do tonight," says Rob. "Put all the stuff in our shoes and go out on the town."

"We'll pretend we're big-time roulette players," Letty says. "We can communicate with each other through meaningful glances."

The four of us get dressed in our gambling outfits, complete with computer sandwiches and battery boats. None of the equipment functions for more than the occasional random buzz. Letty wears dark pants, a blue Oxford cloth shirt, and sumba cloth vest from Bali. I sport the cravat and sports coat of a French restaurateur. Rob, in a Hawaiian shirt opened three buttons at the neck, looks like he just arrived on the last wave from Waikiki. Doyne emerges in white pants and a black shirt. "My mom bought this for me," he says. "When she heard I was going to play roulette in Las Vegas, she wanted me to wear the right kind of clothes. It's a disco suit, which is why it doesn't have a jacket."

We eat a large Mexican dinner of enchiladas washed down with margaritas. We cruise up and down the Strip admiring the neon and then head for the parking lot at Caesars Palace.

Letty turns to Rob. "Are you powered up?"

"My toes are clicking away like crazy," he tells her. "It's the best-fitting pair of shoes I've had in a long time."

"Are you getting any signals?"

"No. Not a thing."

"Good," she says. "Let's go."

Riding the elevated skyway into the casino, we roll past anatomically enhanced centurions and nymphs while listening to a taped message about "the glory that was Rome." The conveyor belt dumps us into a hall filled with slot machines and change ladies, and we make our way from there down a corridor leading to the main gambling floor. On the way we pass various souvenir shops and discotheques, including Cleopatra's Barge, which consists of a wooden structure that looks like a cross between a trireme and a helicopter pad floating in a pool of chlorinated water. Couples dressed in disco suits and party dresses walk up a gangplank to dance on the Barge, where the Bruce Westcott Band is playing a medley of soft core rock.

We push farther down the corridor until it opens onto the main gambling hall. The room is circular, and its domed ceiling twinkles with ersatz starlight. A crowd presses around the tables. Money in the form of silver dollars and chips tumbles over the baize. Hostesses wearing push-up bras, see-through togas, golden crowns, and hairpieces that tumble down their backs like horses' tails circulate with cocktail glasses tinkling on silver trays. It's the Roman Rapunzel look as concocted by Frederick's of Hollywood. The men playing on the floor sport Gucci loafers, pinky rings, and pastel shirts unbuttoned to the navel. The women sashay in spike heels and strapless gowns with cutaway backs, or they wear harem pants tied at the ankle and slit to the thigh. Their hair is whipped into richly teased confections, ratted, frosted, tinted, streaked, and piled high on their heads or frizzed out into Barbra Streisand curls. Leaning back to laugh, the women show their necks to good effect. The men display approval by flashing their teeth and biting off the ends of cigars.

The four of us walk across the floor to stand in the crowd around the roulette tables. Three Asian businessmen playing together in a consortium are winning big. Their hands shake as they finger what must be twenty thousand dollars in chips. Jotting down notes on the back of a postcard, they whisper among themselves. A security guard wheels over a rack of five-hundred-dollar chips, in case the businessmen decide to cash out. Other players buy into the game with large bills that the croupier, using his wooden trowel, stuffs through a slot in the cashbox. Still more players sit along the layout fumbling their chips, adding up the columns in their systems, and otherwise trying to hide the nakedness that comes on being cleaned out by the house.

We watch the roulette wheels spin for an hour. The computers in our shoes are lifeless, but we automatically time the rotors and set parameters in our heads. These wheels are a perfect knockover — nicely tilted and shadowed, with steady rotors and fast balls up on the track. The croupiers couldn't be more docile in offering them up to be beaten.

"Let's go," Doyne says, pulling himself away from the tables. We walk back across the floor and down the corridor to Cleopatra's

Barge, where we climb the gangplank and dance to the soft-core sound of the Bruce Westcott Band. "You saw those wheels back there?" he tells us, with a look of disgust on his face. "We could have killed them"

"You're right," we agree. "We could have killed them."

The Intergalactic Infandibulum

However comical it may be that I should expect to get so
much out of roulette, the routine opinion, accepted by
everybody, that it is absurd and silly to expect anything
at all from gambling seems to me even funnier.

Feodor Dostoyevsky

After a week of storms casting rain and snow over the desert, the sky breaks into patches of blue out of which shines a warm and buttery sun. The peaks of the Sangre de Cristo Mountains sparkle with snow. Down in Santa Fe and nearby Jacona, where a handful of adobe houses straggle along the banks of the Pojoaque River, not far from where it joins the Rio Grande, the first day of spring has laid a carpet of lupine, poppy, mallow, and other desert ephemerals thick underfoot. The fruit and mulberry trees are in bloom. The Russian olives are leafing out along the stream beds. Farther south, toward the red mesa on top of which sits Los Alamos, the jumping cholla and prickly pear are budding alongside white thorn acacias and their sweet-smelling flowers.

In the courtyard of an old adobe house surrounded by Chinese elms, fifty of us — gathered from off the Atlantic and Pacific coasts, or down from the mountains stretched between Idaho and Silver City — stand in a semicircle around Doyne and Letty. Doyne wears a Mexican wedding shirt with a rosebud pinned to his collar. Letty is dressed in a white gown sashed at the waist, with satin panels and lace ties at the sleeves. Standing between them is Dave Miller, former Explorer Scout and New Mexico motocross champion. An engineer turned social worker, Miller presides over this gathering on the banks of the Pojoaque River in his capacity as card-carrying minister in the Universal Life Church. "I'm just here to sign the forms," he says nervously. "This is the first time I've ever done anything like this."

Norman Packard and Letty's sister, Margaretta, stand next to Doyne and Letty as best man and woman. A puckish smile on his face, Norman rifles through his pockets and pretends to have lost the ring box. Everyone laughs when it finally appears. The parents of the bride and groom give their blessing. Minister Miller pronounces the benediction, and then, as everyone presses close around them, he opens a small silver knife and hands it to Letty.

"Almost from the moment we met," says Doyne, in a voice made unsteady with emotion and wind in the courtyard, "Letty and I became best friends, and through everything that's happened to us the closeness and prominence of this friendship has always persisted. Even though we are about to get married, we intend to remain best friends. It seems appropriate, then, that we add to the traditions of matrimony another tradition, long-standing for cementing friendships — and become blood brother and sister as well as husband and wife. So, if Letty can draw some of my blood, without cutting my hand off, and I can similarly draw some of hers, we will interchange a little bit of our blood as a symbol of the closeness of the ties between us. This is also an augury of our becoming a family. We hope that as our blood mingles here, so will it mingle in the veins of our children, who at least in part will be a synthesis of both of us."

There is deep silence during the bloodletting, and tears and smiles when the deed is done. The ceremony finished, we walk to another patio surrounding a pool to eat the wedding dinner of enchiladas made with blue corn tortillas. Late in the afternoon, having imbibed too much champagne and Mexican beer, the bride and groom and remaining guests strip naked to play a spirited game of water polo. At nightfall a dozen of us drive into the Sangre de Cristos on the Taos road. We park in the pine forests above snow line and clamber up the hillside to a spa called Ten Thousand Waves. We undress again and scurry through the snow before plunging up to our necks in a pool of hot water. With steam rising over us to evaporate under a canopy of stars, we talk softly among ourselves and catch up on the news.

Tom Ingerson, still traveling light, is headed back to his observatory in Chile to look for Seyferts in the night skies of the Southern Hemisphere. As much a "synergistic personality" as ever, with the same piercing blue eyes and authoritative voice, he looks the perpetual scout leader, red-cheeked and hearty. Carrying a couple of cotton shirts and a sleeping bag stuffed into a rucksack, the eternal *Wandervogel* is still aspiring to quit academia and exploit his ideas in the company of friends. "I'm trying to circumvent the fissioning pressures of society, the fact that it scatters people and their careers at random. There must be some way, if you can get over the capitalist hump, to gather everyone together and build an organization large enough to subsidize ideas."

Twinned in Ingerson's mind with the idea of founding a company is that of founding a family. Having assigned himself the problem of securing a wife, he is solving it with his customary ratiocination. "I'm a high technocrat," he says. "I work at the cutting edge of technology. But that's not where my greatest loves are. I'm not fond of cities, and I don't much like civilization. I would hate to live always in the world of physics, and if I never programmed another computer, that would be fine with me. For a while, in Silver City, the Explorer Post was my surrogate family. I lived in a big clubhouse. But how could I invite a girl over to dinner when there were disassembled motorcycles on the dining room table? I loved spending time with Norman and Doyne and the other kids, but they bit out of my life the years I normally would have spent settling down to have a family.

"I may be a technocrat, but when I thought about it, I realized my emotional center of gravity was closer to *Mother Earth News*. That's how I came up with the idea of advertising for a wife in their personal columns. Except for their streak of anti-intellectualism. I have a lot of sympathy for the organic people. Living in solar houses and eating healthy food appeals to my basic home-and-hearth instincts. But I wasn't looking for someone who would lecture me on her prejudices about Venus ascending, or tell me that I shouldn't eat eggs on the fifth Tuesday after the summer solstice. So when I sat down to write the advertisement, I tried hard to think of the best way to characterize myself. PHYSICIST sounded too scary. I didn't want a threatening word appearing in boldface. ASTRONOMER was too close to ASTROLOGY. I was looking for something implying a rational world view with no great interest in gods, and the word I came up with was SCIENTIST.

"I took a lot of trouble in the body of the ad describing how I wanted to do three things: build an underground house, sail around the world, and have some kids. I was amazed when I got two hundred seventy-five responses. These were long letters, beginning 'Dear Scientist,' in which people told me their life stories. It was an embarrassment of riches. I was carrying on such a voluminous correspondence that for a while it became the dominant thing in my life. I didn't know where to begin making a choice. I had letters from people all across the country, but I thought maybe I should limit myself to the Pacific Northwest and California and make a little tour."

"And how did it go?" I ask.

"I visited twenty-five respondents, and made some wonderful friends. But for whatever reason, I didn't find the right person. I'm still looking."

As for the other Projectors and their search for Eudaemonia, Jim Crutchfield has carried his computer wizardry to New Mexico. Working with Doyne as hacker-in-residence at the Center for Nonlinear Studies, the two of them have patched together an analog-digital system like the one they had back in Santa Cruz, and this rump group of the Chaos Cabal is hot on the trail to a couple of "breakthrough" ideas in chaos theory.

"I'd be interested in working on the Project again," says Crutchfield. "The technology has advanced so fast that you could build the same computer today with half the chips. That would cut the number of connections way down from the current one hundred and twenty, which would clean up the wiring, the PC boards, and all the rest of it. The physics of the Project is good, but a lot of work has to go into assembling the program and rewriting it in a high-level language, like C. At the moment most of exists in Doyne's head. No one else can make sense out of all those pencil marks. It isn't until everything is shaken out and stared at objectively that the Project can be re-engineered, and for that Doyne needs a total brain dump."

Except for steam rising over us to condense into icicles on the pine trees, the night is crystal clear. I drift along the edge of the pool taking a canvass of Eudaemonic friends. Where have we been? Where are we going? When next will we find ourselves gathered together? Rob Shaw, bearded and jocular, is the last of the Chaos Cabal remaining in Santa Cruz. "Someone had to hold the fort and be a beacon of truth and justice," he says. Still in love with his two great passions, physics and music, Rob has moved all his belongings, including an electric piano, into the physics building, where he is living next to the analog, the NOVA, and his other computers. "If I don't get my grant *this* year," he quips, "I'm going to steal the NOVA and hide it in the trunk of my car. Then they'll see what can be done with a mobile unit."

Grazia Peduzzi is headed back to Italy. The serene republic of Santa Cruz has welcomed her most graciously, but time has come for the traveler to turn homeward. After selling *Star Wars* memorabilia for the toy division of Lucas Films, Ingrid Hoermann, unsung heroine of many a gambing session in Strip and Gulch, is working as a radio engineer in Berkeley. Marianne Walpert, the red-haired bacchante of Riverside Street, is enrolled in the Women in Physics graduate program at Northeastern University. Charlene Peterson and her boyfriend, after saving enough money to buy land in northern California and build a house, have split up. He's doing computer animation; she's into Zen Buddhism. Alix Youmans, the first of the Project's dedicated players, is living in San Diego with a neurosurgeon. Dan Browne, having switched from the physical to the social sciences, is writing a doctoral dissertation on the anthropology of game playing in Japanese ashrams. While doing additional field work in the Oxford Card Room in Missoula, Montana and other favorite haunts in the Pacific Northwest, Browne continues to play a mean game of poker.

Len Zane, after being told by a pit boss at the Sahara that he was stepping out of line, has given up card counting and gone back to running the physics department at the University of Nevada. Bruce Rosenblum, Bill Burke, and George Blumenthal are still keeping an eye on the promising graduate students coming up through the ranks at UC Santa Cruz. Rob Lentz has a new job in electronics — this one having nothing to do with building weapons. Mark Truitt and Wendy Tanizaki have moved to another part of Santa Cruz. She's finishing college. He's looking for work. As Mark described his recent job-hunting experience in a letter to Doyne, "My résumé is not surviving the initial screening process for a variety of reasons. The companies I'm talking to may suspect that 'microcomputer applications at Eudaemonic Enterprises' means playing Pac Man at the Boardwalk arcade. I think recommendations from you and Norman, written on official stationery, might increase my credibility."

Jonathan Kanter is still commuting to the Silicon Valley to sell ideas. Neville Pauli works in San Francisco as an investment banker. Among the early Projectors, John Boyd, having dropped out of graduate school, is living in Seattle. Jack Biles has finished a degree in experimental physics and taken a job at the Oregon Museum of Science and Industry. John Loomis, enrolled in the architecture school at Columbia, lives in New York City. Steve "The Toe" Lawton, balding but otherwise in excellent physical condition, runs a bookstore in Aptos, California, where he keeps the shelves especially well stocked with utopian literature. Alan Lewis, director of research for an investment company in Newport Beach, California, continues to look for ingenious applications of physics to the stock market. Ralph Abraham is writing a series of books illustrated in full color by Chris Shaw. The first of a new genre that he calls "visual mathematics," Abraham's volume on strange attractors is selling briskly. Ranking himself "sixth in the world" among stock market gamblers, Edward Thorp is working on a new, improved system. "I was once the best blackjack player in the world, and I would like to be, for my own satisfaction, the best money manager in the world." He admits to dropping into a casino now and then to practice card counting. "But I'm mainly interested in playing the stock market. It's a much bigger scale thing."

Norman Packard and the Big L, as he fondly refers to Lorna Lyons are still a going concern. They lived together in Europe while Norman worked as a NATO fellow outside Paris. Now, as Norman takes a job at the Institute for Advanced Studies in Princeton, they are transplanting themselves to the East Coast. His latest thinking in chaos theory is about something called "spatial entropy," which he describes as an extension of Claude Shannon's original formula equating information with surprise. Norman is one among a number of Eudaemons for whom computer building in the basement proved excellent training for the more advanced realms of theoretical physics.

Letty Belin, after moving north from Los Angeles to San Francisco, is still practicing law in the public interest. Involved in conservation and class-action cases, she intends to keep her name and career unchanged by marriage. Doyne Farmer, employed at the Center for Nonlinear Studies at the Los Alamos National Laboratory, is busy thinking about what he calls "the information dimension." This is a handy tool for measuring the amount of chaos in a system. Having been named Oppenheimer Fellow at Los Alamos, he now for the first time in his life has enough money to buy some early Dan Hicks records, and the complete Chuck Berry.

Drifting next to him on this cold night in the Sangre de Cristo Mountains, I ask Doyne for his latest thoughts on Eudaemonia. "The Eudaemonic Pie is like air," he replies. "All of us own it according to how much we breathe." He is less cryptic in describing the evolution of the Pie over the lifetime of the Project. "Just like any other pie, it was meant to be cut into slices. The size of your piece was proportional to the time you spent working on the Project. No matter how wonderful your ideas were, it was purely a matter of time put in. Investment capital, taken altogether, was meant to consume a fixed portion of the Pie. But as more capital was required, each dollar bought a smaller and smaller slice of Pie. This happened because, as time went on, and it became clear that the amount of labor was much more than anticipated (to say the least), the investors' slice of the Pie kept getting pared down. There was also initially, due to the insistence of Jack Biles, another section of Pie devoted to 'seminal development,' i.e., the summer Jack and Norman spent in Las Vegas spawning the idea that sent us on this crazy scheme. But this slice of Pie was also substantially trimmed with the passage of time. As it now stands, there are a very large number of people owning very small pieces of Pie, which has gotten lighter and lighter (too much meringue) until it finally started floating in thin air. Then Mark came along and put a 'front end' on it, and the Pie is now orbiting somewhere between here and the intergalactic infandibulum."

Doyne still thinks about the Project and wonders if he could find investors willing to build a new generation of computers. As for his earlier ideas on founding a company and gathering friends together to organize the good life governed according to reason: "It's like climbing a mountain," he observes. "You reach a certain point and get blown away. You go back, thinking you've found a good route. But you don't make it the second time, either, and you're really frustrated, because *that* time you thought you really had found the way to the top."

For all of us floating together under a starry sky, the Eudaemonic Pie is a tangible presence. Conjured out of thin air, it exists wherever Eudaemons gather to talk of it. It assumes a mouth-watering fullness concocted out of stories about the Project, reminiscences, jokes, and plans for building another generation of roulette computers. There is also a deeper understanding shared among us, which involves the realization that Eudaemonia is not a goal to be attained in life, a telos. It is instead a process. We have already known the good life governed according to reason, and it existed for us in the very act of pursuing the Project. Eudaemonia was there all along in the shared experience of living and working together. There had been grander dreams of breaking the bank in Las Vegas to live free of universities and jobs. Of buying land in Washington or Oregon to set up a commune buzzing with appropriate technology. Dreams of travel, of building dirigibles, weightless cubes, cellular automata. But the dreaming itself had partaken of Eudaemonia during the years spent together at 707 Riverside building computers and tripping across the desert to play roulette.

After the Project's final journey to Las Vegas, Mark Truitt sealed the roulette wheel in its shipping crate. Gathering together the KIM, the EPROM burner, the computer sandwiches and boats, the betting practice box, the eye-toe coordination device, and all the magic shoes, he put them into a trunk that he wrapped entirely in black electrical tape. The crated wheel and trunk were buried among a collection of surfboards, wet suits, old TVs, garden tools, motorcycle carburetors, and lumber deep in the nether reaches of the Riverside basement.

No one knows if and when the computer might be retrieved to pay another visit to Caesars Palace, but there is talk of resurrection. In October 1983, six months after the Eudaemonic gathering in Santa Fe, Doyne paid for the following advertisement to run in *Gambling Times:*

INVESTORS
WANTED
for computer
system to beat

roulette using
predictive
physical
principles

This is not a betting scheme.
Small computer predicts
approximate landing point
of ball. Prototype has proven
advantage in casino of
20–40%. Capital needed for
final stages of hardware
development Address
inquiries to:

Giving his address and phone number at Los Alamos, Doyne got a flurry of responses. A lawyer in Miami offered to put up ten thousand dollars. A systems programmer in the Silicon Valley called to talk about how he had tried but failed to build his own roulette computer. He wanted to invest between five and seven thousand dollars, depending on how well his poker playing went in the next few weeks. A fellow named Earl phoned from Las Vegas to describe how he and a group of friends had used concealed computers to work for a number of years as blackjack card counters. The business was lucrative, he said. But with the casinos beginning to catch on to them, Earl and his colleagues needed to branch out into another game. They already had a hundred thousand dollars in capital raised for building a roulette computer.

Another contact acquired from the advertisement in *Gambling Times* was Keith Taft. Taft works out of the Silicon Valley, where he operates what Doyne describes as "a gambling computer supermarket. When I talked to him on the phone, I could hardly believe what he was telling me. Taft specializes in making computer systems for card counters. These are built with toe-operated switches and small machines that are sewn into pouches and strapped to the body. The computers are constructed around Z-80 microprocessors, which are easy enough to reprogram for playing roulette. Getting a concealable computer from Taft is like ordering a take-out pizza. You call him up and say, 'Hi, this is Doyne Farmer in Santa Cruz. I'd like a delivery of three Z-80 microcomputers, four pairs of dingo boots with built-in switches, and three communication systems to go. Hold the blackjack software.'

"He charges six thousand dollars for the software, which makes it the most expensive item on the list. This seems curious to me, since anyone with a PROM burner could borrow one of his friends' computers and bootleg the program. But the rest of his prices are amazingly reasonable. Here's a sampling of what you can buy from his gambler's mail-order house. Item: a Z-80 based microcomputer in an epoxy case with two exposed 2764 PROM sockets (eight thousand bytes each) and two thousand bytes of RAM. This little baby is smaller than a cigarette pack and runs off a size C lithium battery for eleven hours. Price: twenty-five hundred dollars. Item: One pair of dingo boots complete with durable microswitches and solenoids. Price: five hundred dollars. Item: Communication system, consisting of a radio transmitter interfacing to the Z-80 and a passive receiver for driving the shoe-mounted solenoids. Price: a thousand dollars. Systems carry a lifetime guarantee and come complete with computer pouch and wiring harness running from shoe to shoe. Orders filled within two weeks. All major credit cards accepted. Taft is also developing a new CMOS model computer that promises to be smaller and last longer without a change of batteries. But he can't guarantee delivery before the Christmas rush."

Doyne received another phone call from Len Zane in Las Vegas. "Someone just walked into my office," Len said, "and told me he was organizing a team of scientists and investors to build a computer for beating roulette. 'Wait a minute,' I told him. 'I know just the man for you.' The guy's in my office right now jumping up and down with excitement. He wants to throw money at you. He says it's coming from a lady with three million dollars who loves to play the game and wants to invest in a roulette computer just for the pleasure of it."

"It looks like another bout of roulette madness is upon us," Doyne tells me when I last talk to him on the phone. "What do you think about heading back to Las Vegas? Are you ready to take the heat?"

"You can count on me to roast myself in any casino you want. After all, I have a slice of Pie coming to me one of these days."

"I have to admit," Doyne acknowledges, "I'm getting excited again about roulette."

And Afterwards?

Fortunate Newton, happy childhood of science!

Albert Einstein

On May 30, 1985 the Governor of Nevada signed into law Senate Bill 467, which makes it a crime, punishable by up to ten years in prison and a fine of $10,000, to carry into a casino a "device" capable of "projecting the outcome of the game". Does this include pencils and paper? If a croupier hands you a pen for calculating the odds, is he an accessory to a crime? Clearly the casino owners are trying to catch up with computers; they just don't know what to call them.

So far the only Eudaemon to come close to legal harm is Edward Thorp. Thorp's stock market scheme in its heyday grew into a $350 million hedge fund managed from bicoastal offices in Princeton, New Jersey and Newport Beach, California. Princeton/Newport Partners specialized in computerized trades of Eurodollar convertibles, junk bonds, and other instruments with exotic exchange features. "On a conversion operation, if you're thirty seconds too slow," says Thorp, "there's no point in even playing."

Thorp did well enough in the stock market to quit teaching and build a ten-bathroom house on a hilltop overlooking the Pacific Ocean. His partner owned a 225-acre horse farm in New Jersey, but this was apparently inadequate to the man's ambition. After P/NP's offices were raided by fifty federal marshals wearing bulletproof vests, Thorp's partner and four employees—but not Thorp himself—were convicted of insider trading in some of Wall

Street's biggest takeover deals. Princeton/Newport Partners is now out of business.

The wheel of fortune has been kinder to the rest of us. Rob Shaw, for example, after winning a "genius award" of $250,000 from the MacArthur Foundation, is directing The Pocket Institute for the study of chaotic behaviour out of his house in Santa Cruz.

In June 1989 many of the old projectors gathered in Cannobio, Italy, on the shores of Lago Maggiore, for the eudaemonic wedding of Norman Packard and Grazia Peduzzi. The frescoes in the family chapel were newly restored for the occasion. "I would have put the odds on Grazia being married by a priest in the old house at a million to one," said her sister. When last heard from, Grazia was teaching philosophy to high school students in Milan, and Norman, a professor of chaos at the University of Illinois, was designing strange attractors into Italian fabrics by connecting his NeXT computer to the Peduzzi family's weaving machines.

Letty Belin, Doyne Farmer, and their three children still live in New Mexico, where Letty represents the Colorado River Indian tribes, which include the Mojave, Hopi, Navajo, and Chimehuevi. "I blew the opposition out of the water," she says of an Indian land case she won recently in federal court. She is now fighting to block Japanese timber concessions that threaten to level the pine forests in the American Southwest.

Doyne has been promoted to Group Leader in the Theoretical Division at Los Alamos National Laboratory. He makes a point of avoiding research projects with military applications. "I've become a jack of all trades. I'm trying to get the big picture," he says of his work on artificial life, game theory, population biology, and the human immune system—all of which he believes are united by a common chaotic structure.

"I'm thinking of some new projects," he told me recently. "Maybe something to do with creating artificial life. NASA is looking for someone to design self-reproducing aluminium mining modules to be placed on the Moon."

As for roulette, Doyne guesses that about fifteen groups have taken up play since the early days of Eudaemonic Enterprises. One team experimented with video camera inputs the size of shirt buttons. Another person claims to have developed a visual system capable of clocking the wheel mentally. But most teams are still employing the old technology of magic shoes loaded with toe-operated computers. Doyne occasionally receives post cards from exotic places and letters describing big wins under casino heat. "Sure," he says, in response to my hypothetical question, "if somebody invited me to Monte Carlo for the weekend and told me I was going to make $100,000, I could be induced to play roulette again."

FOR THE BEST IN PAPERBACKS, LOOK FOR THE 🐧

In every corner of the world, on every subject under the sun, Penguin represents quality and variety – the very best in publishing today.

For complete information about books available from Penguin – including Puffins, Penguin Classics and Arkana – and how to order them, write to us at the appropriate address below. Please note that for copyright reasons the selection of books varies from country to country.

In the United Kingdom: Please write to *Dept E.P., Penguin Books Ltd, Harmondsworth, Middlesex, UB7 0DA.*

If you have any difficulty in obtaining a title, please send your order with the correct money, plus ten per cent for postage and packaging, to *PO Box No 11, West Drayton, Middlesex*

In the United States: Please write to *Dept BA, Penguin, 299 Murray Hill Parkway, East Rutherford, New Jersey 07073*

In Canada: Please write to *Penguin Books Canada Ltd, 2801 John Street, Markham, Ontario L3R 1B4*

In Australia: Please write to the *Marketing Department, Penguin Books Australia Ltd, P.O. Box 257, Ringwood, Victoria 3134*

In New Zealand: Please write to the *Marketing Department, Penguin Books (NZ) Ltd, Private Bag, Takapuna, Auckland 9*

In India: Please write to *Penguin Overseas Ltd, 706 Eros Apartments, 56 Nehru Place, New Delhi, 110019*

In the Netherlands: Please write to *Penguin Books Netherlands B.V., Postbus 195, NL–1380AD Weesp*

In West Germany: Please write to *Penguin Books Ltd, Friedrichstrasse 10–12, D–6000 Frankfurt/Main 1*

In Spain: Please write to *Alhambra Longman S.A., Fernandez de la Hoz 9, E–28010 Madrid*

In Italy: Please write to *Penguin Italia s.r.l., Via Como 4, I-20096 Pioltello (Milano)*

In France: Please write to *Penguin Books Ltd, 39 Rue de Montmorency, F-75003 Paris*

In Japan: Please write to *Longman Penguin Japan Co Ltd, Yamaguchi Building, 2–12–9 Kanda Jimbocho, Chiyoda-Ku, Tokyo 101*

PENGUIN SCIENCE AND MATHEMATICS

QED   Richard Feynman
The Strange Theory of Light and Matter

Quantum thermodynamics – or QED for short – is the 'strange theory' – that explains how light and electrons interact. 'Physics Nobelist Feynman simply cannot help being original. In this quirky, fascinating book, he explains to laymen the quantum theory of light – a theory to which he made decisive contributions' – *New Yorker*

God and the New Physics   Paul Davies

Can science, now come of age, offer a surer path to God than religion? This 'very interesting' (*New Scientist*) book suggests it can.

Does God Play Dice?   Ian Stewart
The New Mathematics of Chaos

To cope with the truth of a chaotic world, pioneering mathematicians have developed chaos theory. *Does God Play Dice?* makes accessible the basic principles and many practical applications of one of the most extraordinary – and mindbending – breakthroughs in recent years. 'Engaging, accurate and accessible to the uninitiated' – *Nature*

The Blind Watchmaker   Richard Dawkins

'An enchantingly witty and persuasive neo-Darwinist attack on the anti-evolutionists, pleasurably intelligible to the scientifically illiterate' – Hermione Lee in the *Observer* Books of the Year

The Making of the Atomic Bomb   Richard Rhodes

'Rhodes handles his rich trove of material with the skill of a master novelist ... his portraits of the leading figures are three-dimensional and penetrating ... the sheer momentum of the narrative is breathtaking ... a book to read and to read again' – Walter C. Patterson in the *Guardian*

Asimov's New Guide to Science   Isaac Asimov

A classic work brought up to date – far and away the best one-volume survey of all the physical and biological sciences.

A CHOICE OF PENGUINS

The Secret Lives of Trebitsch Lincoln Bernard Wasserstein

Trebitsch Lincoln was Member of Parliament, international spy, right-wing revolutionary, Buddhist monk – and this century's most extraordinary conman. 'Surely the final work on a truly extraordinary career' – Hugh Trevor-Roper. 'An utterly improbable story ... a biographical coup' – *Guardian*

Out of Africa Karen Blixen (Isak Dinesen)

After the failure of her coffee-farm in Kenya, where she lived from 1913 to 1931, Karen Blixen went home to Denmark and wrote this unforgettable account of her experiences. 'No reader can put the book down without some share in the author's poignant farewell to her farm' – *Observer*

In My Wildest Dreams Leslie Thomas

The autobiography of Leslie Thomas, author of *The Magic Army* and *The Dearest and the Best*. From Barnardo boy to original virgin soldier, from apprentice journalist to famous novelist, it is an amazing story. 'Hugely enjoyable' – *Daily Express*

The Winning Streak Walter Goldsmith and David Clutterbuck

Marks and Spencer, Saatchi and Saatchi, United Biscuits, GEC ... The UK's top companies reveal their formulas for success, in an important and stimulating book that no British manager can afford to ignore.

Bird of Life, Bird of Death Jonathan Evan Maslow

In the summer of 1983 Jonathan Maslow set out to find the quetzal. In doing so, he placed himself between the natural and unnatural histories of Central America, between the vulnerable magnificence of nature and the terrible destructiveness of man. 'A wonderful book' – *The New York Times Book Review*

Mob Star Gene Mustain and Jerry Capeci

Handsome, charming, deadly, John Gotti is the real-life Mafia boss at the head of New York's most feared criminal family. *Mob Star* tells the chilling and compelling story of the rise to power of the most powerful criminal in America.

A CHOICE OF PENGUINS

The Assassination of Federico García Lorca   Ian Gibson

Lorca's 'crime' was his antipathy to pomposity, conformity and intolerance. His punishment was murder. Ian Gibson – author of the acclaimed new biography of Lorca – reveals the truth about his death and the atmosphere in Spain that allowed it to happen.

Between the Woods and the Water   Patrick Leigh Fermor

Patrick Leigh Fermor continues his celebrated account – begun in *A Time of Gifts* – of his journey on foot from the Hook of Holland to Constantinople. 'Even better than everyone says it is' – Peter Levi. 'Indescribably rich and beautiful' – *Guardian*

The Time Out Film Guide   Edited by Tom Milne

The definitive, up-to-the-minute directory of 9,000 films – world cinema from classics and silent epics to reissues and the latest releases – assessed by two decades of *Time Out* reviewers. 'In my opinion the best and most comprehensive' – Barry Norman

Metamagical Themas   Douglas R. Hofstadter

This astonishing sequel to the bestselling, Pulitzer Prize-winning *Gödel, Escher, Bach* swarms with 'extraordinary ideas, brilliant fables, deep philosophical questions and Carrollian word play' – Martin Gardner

Into the Heart of Borneo   Redmond O'Hanlon

'Perceptive, hilarious and at the same time a serious natural-history journey into one of the last remaining unspoilt paradises' – *New Statesman*. 'Consistently exciting, often funny and erudite without ever being overwhelming' – *Punch*

When the Wind Blows   Raymond Briggs

'A visual parable against nuclear war: all the more chilling for being in the form of a strip cartoon' – *Sunday Times*. 'The most eloquent anti-Bomb statement you are likely to read' – *Daily Mail*

A CHOICE OF PENGUINS

Brian Epstein: The Man Who Made the Beatles   Ray Coleman

'An excellent biography of Brian Epstein, the lonely, gifted man whose artistic faith and bond with the Beatles never wavered – and whose recognition of genius created a cultural era, even though it destroyed him' – *Mail on Sunday*

A Thief in the Night   John Cornwell

A veil of suspicion and secrecy surrounds the last hours of Pope John Paul I, whose thirty-three day reign ended in a reported heart attack on the night of 28 September 1978. Award-winning crime writer John Cornwell was invited by the Vatican to investigate. 'The best detective story you will ever read' – *Daily Mail*

Among the Russians   Colin Thubron

One man's solitary journey by car across Russia provides an enthralling and revealing account of the habits and idiosyncrasies of a fascinating people. 'He sees things with the freshness of an innocent and the erudition of a scholar' – *Daily Telegraph*

Higher than Hope   Fatima Meer

The authorized biography of Nelson Mandela. 'An astonishing read ... the most complete, authoritative and moving tribute thus far' – *Time Out*

Stones of Aran: Pilgrimage   Tim Robinson

Arainn is the largest of the three Aran Islands, and one of the world's oldest landscapes. This 'wholly irresistible' (*Observer*) and uncategorizable book charts a sunwise journey around its coast – and explores an open secret, teasing out the paradoxes of a terrain at once bare and densely inscribed.

Bernard Shaw   Michael Holroyd
Volume I 1856–1898: The Search for Love

'In every sense, a spectacular piece of work ... A feat of style as much as of research, which will surely make it a flamboyant new landmark in modern English life-writing' – Richard Holmes in *The Times*

FOR THE BEST IN PAPERBACKS, LOOK FOR THE

A CHOICE OF PENGUINS

The Russian Album   Michael Ignatieff

Michael Ignatieff movingly comes to terms with the meaning of his own family's memories and histories, in a book that is both an extraordinary account of the search for roots and a dramatic and poignant chronicle of four generations of a Russian family.

Beyond the Blue Horizon   Alexander Frater

The romance and excitement of the legendary Imperial Airways East-bound Empire service – the world's longest and most adventurous scheduled air route – relived fifty years later in one of the most original travel books of the decade. 'The find of the year' – *Today*

Getting to Know the General   Graham Greene

'In August 1981 my bag was packed for my fifth visit to Panama when the news came to me over the telephone of the death of General Omar Torrijos Herrera, my friend and host...' 'Vigorous, deeply felt, at times funny, and for Greene surprisingly frank' – *Sunday Times*

The Time of My Life   Denis Healey

'Denis Healey's memoirs have been rightly hailed for their intelligence, wit and charm ... *The Time of My Life* should be read, certainly for pleasure, but also for profit ... he bestrides the post-war world, a Colossus of a kind' – *Independent*

Arabian Sands   Wilfred Thesiger

'In the tradition of Burton, Doughty, Lawrence, Philby and Thomas, it is, very likely, the book about Arabia to end all books about Arabia' – *Daily Telegraph*

Adieux: A Farewell to Sartre   Simone de Beauvoir

A devastatingly frank account of the last years of Sartre's life, and his death, by the woman who for more than half a century shared that life. 'A true labour of love, there is about it a touching sadness, a mingling of the personal with the impersonal and timeless which Sartre himself would surely have liked and understood' – *Listener*

FOR THE BEST IN PAPERBACKS, LOOK FOR THE

A CHOICE OF PENGUINS

Riding the Iron Rooster Paul Theroux

An eye-opening and entertaining account of travels in old and new China, from the author of *The Great Railway Bazaar*. 'Mr Theroux cannot write badly ... in the course of a year there was almost no train in the vast Chinese rail network on which he did not travel' – Ludovic Kennedy

The Life of Graham Greene Norman Sherry
Volume One 1904–1939

'Probably the best biography ever of a living author' – Philip French in the *Listener*. Graham Greene has always maintained a discreet distance from his reading public. This volume reconstructs his first thirty-five years to create one of the most revealing literary biographies of the decade.

The Chinese David Bonavia

'I can think of no other work which so urbanely and entertainingly succeeds in introducing the general Western reader to China' – *Sunday Telegraph*

All the Wrong Places James Fenton

Who else but James Fenton could have played a Bach prelude on the presidential piano – and stolen one of Imelda's towels – on the very day Marcos left his palace in Manila? 'He is the most professional of amateur war correspondents, a true though unusual journo, top of the trade. When he arrives in town, prudent dictators pack their bags and quit' – *The Times*

Voices of the Old Sea Norman Lewis

'Limpidly and lovingly, Norman Lewis has caught the helpless, unwitting, often foolish, but always hopeful village in its dying summers, and saved the tragedy with sublime comedy' – *Observer*

Ninety-Two Days Evelyn Waugh

With characteristic honesty, Evelyn Waugh here debunks the romantic notions attached to rough travelling. His journey in Guiana and Brazil is difficult, dangerous and extremely uncomfortable, and his account of it is witty and unquestionably compelling.

FOR THE BEST IN PAPERBACKS, LOOK FOR THE

A CHOICE OF PENGUINS

Return to the Marshes Gavin Young

His remarkable portrait of the remote and beautiful world of the Marsh Arabs, whose centuries-old existence is now threatened with extinction by twentieth-century warfare.

The Big Red Train Ride Eric Newby

From Moscow to the Pacific on the Trans-Siberian Railway is an eight-day journey of nearly six thousand miles through seven time zones. In 1977 Eric Newby set out with his wife, an official guide and a photographer on this journey.

Warhol Victor Bockris

'This is the kind of book I like: it tells me the things I want to know about the artist, what he ate, what he wore, who he knew (in his case ... everybody), at what time he went to bed and with whom, and, most important of all, his work habits' – *Independent*

1001 Ways to Save the Planet Bernadette Vallely

There are 1001 changes that *everyone* can make in their lives *today* to bring about a greener environment – whether at home or at work, on holiday or away on business. Action that you can take *now*, and that you won't find too difficult to take. This practical guide shows you how.

Bitter Fame Anne Stevenson
A Life of Sylvia Plath

'A sobering and salutary attempt to estimate what Plath was, what she achieved and what it cost her ... This is the only portrait which answers Ted Hughes's image of the poet as Ariel, not the ethereal bright pure roving sprite, but Ariel trapped in Prospero's pine and raging to be free' – *Sunday Telegraph*

The Venetian Empire Jan Morris

For six centuries the Republic of Venice was a maritime empire of coasts, islands and fortresses. Jan Morris reconstructs this glittering dominion in the form of a sea voyage along the historic Venetian trade routes from Venice itself to Greece, Crete and Cyprus.

A CHOICE OF PENGUINS

Trail of Havoc Patrick Marnham

The murder of the 7th Earl of Lucan's nanny at the family's Belgravia mansion in 1974 remains one of the most celebrated mysteries in British criminal history. In this brilliant piece of detective work Patrick Marnham investigates the Lucan case and its implications and arrives at some surprising conclusions: what was not disclosed at the murder inquest and what probably *did* happen on that fateful night.

Daddy, We Hardly Knew You Germaine Greer

'It's part biography, part travelogue, its author obsessively scouring three continents for clues to her dead father's identity ... ruthlessly stripping away the ornate masks with which [he] hid his own flawed humanity' – *Time Out*. 'Remarkable, beautifully written' – Anthony Storr

Reports from the holocaust Larry Kramer

'Larry Kramer is one of America's most valuable troublemakers. I hope he never lowers his voice' – Susan Sontag. It was Larry Kramer who first fought to make America aware – notably in his play *The Normal Heart* – of the scope of the AIDS epidemic. 'More than a political autobiography, *Reports* is an indictment of a world that allows AIDS to continue' – *Newsday*

A Far Cry Mary Benson

'A remarkable life, bravely lived ... A lovely book and a piece of history' – Nadine Gordimer. 'One of those rare autobiographies which can tell a moving personal story and illuminate a public political drama. It recounts the South African battles against apartheid with a new freshness and intimacy' – *Observer*

The Fate of the Forest Susanna Hecht and Alexander Cockburn

In a panorama that encompasses history, ecology, botany and economics, *The Fate of the Forest* tells the story of the delusions and greed that have shaped the Amazon's history – and shows how it can be saved. 'This discriminating and constructive book is a must' – *Sunday Times*